A PRETEEN SPEAKS ON
SCIENCE AND TECHNOLOGY
A GUIDE TO YOUNG SCIENTISTS AND TECHNOLOGISTS

TIMOTHY PAUL

INDIA • SINGAPORE • MALAYSIA

Notion Press

Old No. 38, New No. 6
McNichols Road, Chetpet
Chennai - 600 031

First Published by Notion Press 2019
Copyright © Timothy Paul 2019
All Rights Reserved.

ISBN 978-1-64733-613-4

This book has been published with all efforts taken to make the material error-free after the consent of the author. However, the author and the publisher do not assume and hereby disclaim any liability to any party for any loss, damage, or disruption caused by errors or omissions, whether such errors or omissions result from negligence, accident, or any other cause.

While every effort has been made to avoid any mistake or omission, this publication is being sold on the condition and understanding that neither the author nor the publishers or printers would be liable in any manner to any person by reason of any mistake or omission in this publication or for any action taken or omitted to be taken or advice rendered or accepted on the basis of this work. For any defect in printing or binding the publishers will be liable only to replace the defective copy by another copy of this work then available.

Contents

Acknowledgments	7
Foreword	9
SCIENCE — GENERAL	**11**
1. Ten of Mankind's Greatest Inventions	13
2. Nine Misconceptions About the Electronic Devices We Use Every Day	19
3. Five Alternative Materials to Plastic	23
4. Amazing Facts That Will Teach You Something New	28
5. How Can We Resurrect Extinct Animals?	31
ASTRONOMY	**35**
6. Ten Unique Planets You Won't Believe Exist	37
7. What is the Hottest Object in the Universe?	40
AERONAUTICS	**43**
8. How Do Airplanes Fly?	45
9. How Do Helicopters Fly?	49
AUTOMOBILE	**53**
10. Mankind's Quest for Speed	55
11. Automobile Industry	58

ATOMIC SCIENCE — 61
12. Nuclear Energy — 63

SPACE SCIENCE — 69
13. The Achievements of NASA — 71
14. NASA's Plan to Return to the Moon and Stay — 76

THERMODYNAMICS — 79
15. Absolute Zero — 81

NEUROSCIENCE — 83
16. How Does Our Brain Store Memories? — 85

MODERN PHYSICS — 87
17. Higher Dimensions — 89
18. Antimatter — 91
19. The Story of a Genius: Albert Einstein — 93

ENVIRONMENTAL SCIENCE — 97
20. Global Warming — 99

HEALTH SCIENCE — 103
21. The World Health Organization (WHO) — 105
22. Fast Food or Healthy Food – What Should We Eat? — 108
23. Does Advertising Influence the Food Choices of Children? — 111

COMPUTER SCIENCE — 113
24. Parts of a Computer and Their Functions — 115

25.	Lifespan of a PC	118
26.	Interesting Facts About Computers	120
27.	Computers and Technology – Questions and Answers	122
28.	Artificial Intelligence (AI)	126
29.	What Have People Gained by Watching and Making Tech YouTube Videos?	135
30.	How to Make YouTube Videos	137
31.	How Do Websites Store Your Password?	141
32.	Copyrighted Numbers	145
33.	The Most Influential Innovators and Their History	147
34.	The History of Microsoft	152
35.	The Most Innovative Software and Websites	155
36.	Cloud Computing	159
37.	Computers and Gaming in Today's Modern World	162
38.	The History of Video Games	165
39.	Video Games – Good or Not?	167
40.	Watching TV or Playing Video Games by Kids – Pros and Cons	169
41.	The Huawei Ban Explained	171

MATHEMATICS   **173**

42.	Pi (π – The Magical Number)	175
43.	Why Can't We Divide Any Number by Zero?	177

ENGLISH FOR SCIENCE AND TECHNOLOGY 181

44. The Impact of Science and Technology YouTube Videos and English TV Programs on My Speaking Skills – Part I (My Journey Through the World of English)   183

45. The Impact of Science and Technology YouTube Videos and English TV Programs on My Speaking Skills – Part II   188

References   *193*

Links for Pictures   *201*

Acknowledgments

First, I would like to thank my Lord and Savior Jesus Christ for endowing me with intelligence and providing me with an opportunity to write this book. I thank my parents, Dr. Isaac Jebastine and Dr. Lydia Arulselvi, for their constant support, love and encouragement. They were instrumental in making this book. I also thank my uncle Dr. Gibson for giving me the idea and inspiring me to write this book. Finally, I thank my grandparents, Mr. Chelladurai Devaraj and Dr. Vedamani Balraj and Mrs. Mercy Balraj, for their support and prayers.

Foreword

Dr. R. Gokulakrishnan
Additional Director
Software Technology
Parks of India
Ministry of Electronics and
Information Technology
Government of India

No. 5, Third Floor
Rajiv Gandhi Salai
Tharamani
Chennai - 600113

A young author, computer prodigy and YouTuber Timothy Paul's "A Preteen Speaks on Science and Technology " is a welcome addition to the existing body of knowledge of science and technology.

Though the author Timothy Paul is 11 years old, his knowledge in various branches of science, especially how technology influences space science, nuclear science, computer science, automobiles, and so on, and his succinct presentation of scientific facts and technological developments prove that he possesses the intelligence and skills of an adult writer.

The author's cognitive ability is reflected in the treatment of his topics. The many authentic references to YouTube videos and articles in the book demonstrate the author's vast watching and reading.

My watching the author's videos on his YouTube channel Frentran has established the fact that he is also a good and gifted speaker who uses a vast and apt vocabulary which includes scientific terms.

The young author Timothy Paul is able to apply his acquired skills in technology to address real-life issues. And it is my belief that he is capable of bringing laurels to India by winning the prestigious Nobel Prize or other highest international awards.

The author has eschewed unnecessary details, and the presentation is, on the whole, perspicacious. And it is my hope that this book will be well received by the students, especially by young scientists and technologists.

Dr. R. Gokulakrishnan

SCIENCE — GENERAL

1
Ten of Mankind's Greatest Inventions

Here are ten of mankind's greatest inventions.

No: 10 – Gas Stove

The gas stove has certainly revolutionized the way we cook. We do not use open fire anymore. We do not need to chop down firewood also. In the olden days, in most of the Indian houses, cooking was done using firewood. But, nowadays, most people, including chefs, prefer gas stoves. Gas stove was invented and patented by James Sharp in England in 1826.

No: 9 – Paper

Paper is a revolutionary invention. We write everything on paper—documents, tests, sums and even the great inventions. We also print on paper. Paper has a good story behind it.

The invention of paper is commonly credited to Ts'ai Lun (Cai Lun). He ground mulberry barks and hemp rags together with water, then pressed the mixture and dried it on sheets with wooden frames. **(1.1)** We do not use mulberry barks and strands of hemp anymore.

We just use wood pulp. This method was created by an entomologist Rene Antoine Ferchault de Reaumur. He got this idea from observing some wasps that were building a nest. They were munching on wood instead of swallowing it. Then they spat it out and dried. Thus, they formed their nests with paper. Seeing this, he developed the following formula:

> wood pulp + energy + water = paper

This formula is used even today to make the modern kind of paper that we use and love. **(1.2)**

No: 8 – Television

Television is a wonderful modern invention that allows us to hear news and watch cartoons and matches. The electronic television was invented by Philo Taylor Farnsworth in 1927. Many people theorized the television possibilities before Taylor, but he was the first one to prove that images could be displayed on the screen and demonstrated it in front of a crowd. He showed that moving images could be captured using a beam of electrons. Now there are many display types such as LED (light-emitting diode), LCD (liquid crystal display), OLED (organic light-emitting diode) and AMOLED (active-matrix organic light-emitting diode).

No: 7 – Washing Machine

The electrically powered washing machine was invented and designed by Hurley Machine Company engineer Alva John Fisher. Nowadays, we do not have to beat the clothes ruggedly, put soap on them and rub them to remove dirt. A machine does the whole washing for us, and it is an important invention. Now we just throw up our clothes in and close the door, turn up a dial and put some soap. It washes the clothes, rinses and turns them around, dries them and our clothes come out clean. That is how amazing a washing machine is.

No: 6 – Wheel

This object got us moving! No single person or a group can be credited to the making of the wheel. We just know that early humans invented the wheel. As of now, we do not have any proof or evidence for a single person introducing it.

But the modern type, the one we use now, was invented by John Boyd Dunlop, a veterinary surgeon. As he was familiar with making rubber devices, he reinvented pneumatic tires for his child's tricycle and then developed them for use in cycle racing. He sold his rights of the pneumatic tires to a company he formed with Harvey Du Cros, the president of the Irish Cyclists' Association. **(1.3)** The company was first called "Du Cross Creation" and the name was later changed to "Dunlop Pneumatic Tyre Company" after Dunlop became famous.

No: 5 – Smartphone

This amazing modern marvel connects us to the internet and allows us to carry encyclopedias in our pockets. This may be the eighth wonder of the world because it is amazing. We are proudly carrying it in our pockets. It is also known as "Pocket Computer" and it almost matches the computing power of a computer. The first model was IBM (International Business Machines) Simon Personal Communicator which was a handheld, touchscreen smartphone. It first went on sale to the public on August 16, 1994. It took on the term "Smartphone" in 1995. **(1.4)**

No: 4 – Microwave Oven

Another revolutionary cooking device is the microwave oven. Many conspire against it and call it very unsafe, but it has been proven to be safe.

Before the microwave oven was invented, if you had to bake a cake or a pastry or any dessert, you had to put it in a fire oven and evenly distribute the heat. But, now you do not have to. You keep the uncooked food in, click a few buttons, set the time and there you go! You have the deliciously cooked meal ready for you. The microwave oven was invented by the American engineer Percy Spencer in 1946. As you know, the microwave oven runs on electricity.

No: 3 – Electricity

Electricity is an invention that has transformed our lives and the world. Almost everything and anything that we see or use today runs on electricity. It has become a part of our modern lives and we cannot think of a world without it. We need electricity for lighting, charging our smartphones, running our computers, microwave ovens and washing machines, and so on. What can we do without this amazing form of energy? Electricity is the most useful form of energy in the world. You can convert electrical energy into light energy, heat energy, mechanical energy and so on. There are many ways to produce electricity. Different sources of electricity can be divided into two categories: renewable and non-renewable. We owe a great deal to this wonderful form of energy. It is widely debated as to who discovered it.

No: 2 – Computer

Now comes the modern marvel that electricity powers. That is the computer.

Computer is an important and indispensable device in our lives and even the smartphone is just a tiny computer. We have to thank Charles Babbage for making the computer. It is just a string of 1s and 0s. We can connect it to the internet, access media and play video games. We have many encyclopedias on it. It has endless uses. This is one of the most amazing machines that mankind has ever produced.

No: 1 – Internet

The initial idea of the internet is credited to Leonard Kleinrock. The ideas of Kleinrock and Licklider helped Robert Taylor conceive the idea of the network. It was first known as ARPANET and was used in the military. Then it was released to the public and many people started using it. In the 1990s, it was booming. The internet is considered one of the most important things ever created by and for man. There is so much to do with the internet. You can access huge databases with the internet facility. You have the world in your hand or on your desk because of the internet. It connects many things and it connects you from one place to the other. Through internet calls, you can call from America to India for free. It is an amazing invention.

2

Nine Misconceptions About the Electronic Devices We Use Every Day

I am going to state nine misconceptions among people about electronic devices and subsequently discuss the reasons why they are only myths.

Myth 1: You cannot take good photos unless your camera has more megapixels.

Truth: It is one of the most commonly held beliefs that is not true at all. Good image quality is determined not only by having more megapixels but also by many other factors. As *The New York Times* states, "A camera's lens circuitry and sensor and the camera's controls are far more important factors than megapixels." **(2.1)**

Myth 2: If you use a cell phone at a gas station/petrol bunk, it will be a fire hazard.

Truth: Only one gas station or petrol bunk fire has reportedly involved a cell phone, and even in that case,

it was later found that the phone was not the cause of the fire. In our daily news reports, several studies have also propelled the long-time myth that cell phones might generate an electric spark that could ignite gasoline or petrol or other fossil fuels. Just to be safe, the National Fire Protection Association (NFPA) advises you to follow the manufacturer's instructions that you should leave your phone in the car. The Federal Communication Commission (FCC) points out that the fire caused by wireless devices is very remote. **(2.2)**

Myth 3: The World Wide Web (WWW) and the internet are the same.

Truth: They are not the same. The internet is the infrastructure that allows information to be shared between networks across the world including the ones we access via PCs, smartphones and various kinds of software. The web or WWW is one of those networks and it has many sites and pages.

Myth 4: A magnet will erase your data.

Truth: You need a big magnet but even then, it can erase only certain types of data storage, such as solid state drives (SSD) and hard drives—the ones used in your computer. Only powerful magnets like the ones that are used in MRI (magnetic resonance imaging) machines have the power to erase data.

Myth 5: Using a microwave oven is not good.

Truth: While microwave ovens sometimes do leak an amount of radiation, usually nowhere near enough to do any real harm at all. *The New York Times* reports:

"In fact the Food and Drug Administration (FDA) puts limits on amount of microwave's radiation that can leave from an appliance over its lifetime. But, the amount of radiation leaked is far below the level known to harm people." **(2.3)**

Myth 6: You should never put metal implements in the microwave oven.

Truth: It is always not true. Putting metal objects in the microwave oven can be dangerous and as you can see in the picture, 'e' represents the electrons and the electrons are spread in the spoon but the electrons in the fork are concentrated on the points. **(2.4)** They build up and spark but it is not as harmful as you think. Sharp edges can conduct electrons resulting in sparking, but implements with more rounded metal surfaces such as spoons usually do not do any harm.

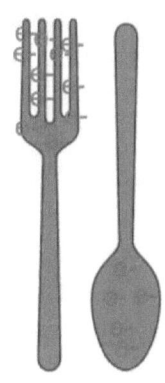

Myth 7: Private or incognito mode means you are anonymous online.

Truth: Enabling private or incognito mode may hide the site that you visit from your web history, but it does not hide you from the web entirely. It means that the site you visit will not go into your web history or automatically log you into your account. It does not mean that you are invisible to the sites you visit.

Myth 8: Cell phones cause brain cancer.

Truth: It was one of the most talked healthcare topics from the 1990s to the early 2000s. No research has found any link between using cell phones and getting cancer.

Myth 9: Playing video games turns good kids into bad ones.

Truth: While video games have been often blamed for kids' bad behavior, it is not the full story. Looking at data spread over ten years, a 2013 study from the University of Glasgow says playing video games consistently does not significantly alter behavior. **(2.5)** In fact, a separate study revealed that kids who played video games for less than an hour were a bit happier and well-adjusted than kids who did not play video games at all. Other studies show that playing video games can have a negative impact, particularly when a game is hard. But studies like these have been disproved.

3
Five Alternative Materials to Plastic

What are the best alternatives to plastic? How can we save the living creatures on the earth and our environment from the hazards of plastic?

First of all, when was plastic invented? Somewhere around 1907, plastic was invented by making chains of polymer not found in the environment. So, microorganisms do not know how to break down the chains of polymer as they do with paper. Microorganisms are able to break down the paper because it is made up of wood pulp which is found in the environment. We use plastic because it is cheap. It has many negatives, in the sense that it pollutes the oceans, the soil and the water, but it also has a very small positive side. It is cheap and durable, but it takes about 500–1000 years to properly disintegrate, but even then, it does not biodegrade. It partially breaks down and turns into microplastics. Microplastics are found in toothpastes, shower gels and shampoos. So, you would not want to swallow your toothpaste next time!

Microplastics are just as harmful as plastics. When you throw plastics into the ocean, due to UV radiation and exposure to the ocean, they just turn into microplastics. Some marine animals eat plastic and some do not. Why? When they become microplastics, they are smaller than 5 millimeters with vibrant color. Marine animals think they are food. They munch them up and now you can guess what happens. They die because they do not know how toxic they are. They contain several harmful things that can cause all sorts of diseases including tumors and cancer. Then why are we still using plastic? And why are we throwing plastic in the oceans? Well, we presume that the use of plastic is not detrimental to our health. But, unfortunately, it does affect us. Sea planktons, which are small marine animals, eat the microplastics. These sea planktons are eaten by small fishes, the small fishes by medium-sized fishes and the medium-sized fishes by us. So, we are indirectly eating microplastics and plastics. **(3.1)** Knowing this, we still are littering. If we continue this, there is going to be no end. We should stop using plastic disposables and just turn to better alternatives.

Edible cutlery is the first replacement for plastic. This is a set of edible cutlery made by a company called Bakey's. Bakey's was founded in 2011 in Hyderabad, India. The cutlery is made with wheat, rice and millets. And this makes for a healthy nutritious spoon and also a fork that you can eat with. This is a revolutionary idea. **(3.2)** So, you could use it and also eat it as it is very nutritious. In addition, it

Image similar to product

contains no preservatives. This can help solve the plastic crisis by eliminating plastic spoons and forks. The edible cutlery has a shelf life of three years. The edible spoons are just millet biscuits. One hundred and twenty billion plastic spoons, forks and plates are thrown or disposed of in India every year, and this alternative could help change the scenario. This is a small but effective step.

The second alternative to plastic is a bottle made from algae, and it is completely biodegradable and edible. This water bottle disintegrates once the purpose is served. This bottle is called "agari" and it is made out of agar which is a jelly-like substance obtained from red algae. **(3.3)** This material is 100% natural and 100% biodegradable. The bottle keeps its shape until it is empty. Then it begins to decompose. You can either throw it away or eat it. It was created by Icelandic product designer Ari Jónsson.

India is in a huge plastic disposal crisis. Huge plastic production, lack of proper disposal mechanism and shortage of funds have led to this huge plastic crisis. The plastic disposal services have failed to keep up with the growth. The third alternative is a bag made from a plant called "cassava." **(3.4)** This bag is biodegradable and looks just like a plastic bag. Though it is twice as expensive as its plastic counterpart, it can save the environment. It is

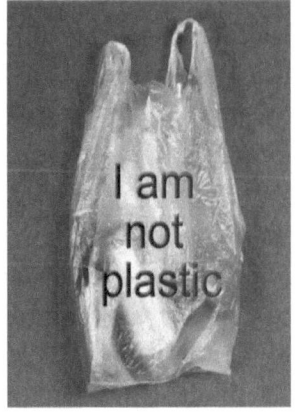

biodegradable; moreover, it biodegrades in about three to four months. This cannot be compared to plastic that takes thousands of years to disintegrate and even then, it does not disintegrate completely. So, this is a welcome change.

The fourth alternative is 3D printing with mycelium and fungus. This opens up a whole new array of possibilities because you can use this for almost anything. This is a 3D-printed material made from fungus. Tiny branches or filaments grow from the core of the fungus and they keep on growing until you want them to stop growing. And how do you stop the growth of a branch? You cook it. And when it grows you could choose its consistency. If you want it to be leathery, you leave it for long; if you want it to be flexible and rubbery, you leave it for a medium amount of time; and if you want it to be soft and delicate, you can leave it around for a short amount of time. **(3.5)** This is a very flexible material. A person named Eric Klarenbeek has even made a chair with mycelium, and you can see mushrooms growing out of it. It is a living material and I could definitely see a future with objects made from mycelium everywhere. There is a company already using it for packaging. Some have started a company that makes mycelium products for everyone. We could ditch the plastic and adopt this as an alternative to plastic. This is not edible because it is made out of fungus. This

amazing idea could just change the way we live and this is definitely going to replace plastic.

The fifth alternative to plastic is an edible sphere of water. It is called Ooho. Ooho is an edible jelly-like sphere filled with drinkable liquid. It is made out of natural, biodegradable ingredients. They decompose after 4–6 weeks if not consumed. The company Skipping Rock Labs that made it spent three years developing Ooho. The company produces about 2000 Oohos a day. **(3.6)**

These are the five alternatives to plastic. We should reduce the use of plastic disposables and should not throw them on the land or in the ocean because now we know that plastic does affect us indirectly. We do not want to eat plastic! Think twice before you put a plastic bag or any other plastic item down. Let us just wait until these alternatives to plastic come to the market exceedingly. Till then, we could use organic materials. It's time we made the earth a better place to live in.

4
Amazing Facts That Will Teach You Something New

Do you know?

1. In 1889, the queen of Italy, Margherita of Savoy, ordered the first pizza delivery.

2. You can buy eel-flavored ice cream in Japan which features eel extracts from an eel's organs.

3. Gecko feet have millions of tiny hairs that stick to surfaces, and the foot structure enables the gecko to climb on walls or trees and hang on by just one toe.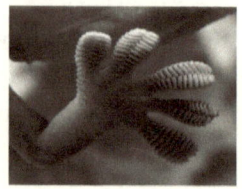

4. The term "Astronaut" comes from two Greek words that mean star and sailor. Star is *Asteri* and sailor is *Naftis*, making *Asteri* into Astro and *Naftis* into Naut, thus making it *Astronaftis*, the sailor of the stars.

5. It is considered rude to write and correct test papers in red ink in Portugal because it is discouraging for students. So, it is banned from use.

6. A cat's tail contains nearly 10% of all the bones in its body for intricate movements.

7. The Nile crocodile can hold its breath underwater for up to two hours while waiting for its prey.

8. Jellyfish or jellies, as scientists call them, are not fish. They have no brain, no heart and no bones. Overall, they do not have any organs.

9. A group of jellyfish is not a herd or a school or a flock; it is called a "smack."

10. The Chinese Giant Salamander can grow to be six feet long, making it the largest amphibian in the world.

11. People reportedly prefer blue toothbrushes to red ones.

12. Scientists say that the best time to take a nap is between 1 p.m. and 2:30 p.m. because that is when a dip in body temperature makes us feel sleepy.

13. The speed of Earth's rotation changes over time. So, a day in the age of Dinosaurs was just 23 hours long as opposed to the current 24 hours.

14. A hummingbird's wings can beat up to 200 times a second.

15. A seahorse can move its eyes in opposite directions in order to scan the water for food and predators.

16. It would take a hundred Earths lined up end-to-end to stretch across the face of the Sun.

17. Some apples can weigh about as much as a one-and-a-half liter of milk.

18. Armadillo is a nocturnal insectivorous mammal and in Spanish "Armadillo" means little armored one.

(Source: 4.1)

5
How Can We Resurrect Extinct Animals?

Can we bring extinct animals back to life? Is it a fantasy? Or is it scientifically true? First, we need to know what extinction is. Extinction means an animal species or breed that totally dies out due to weather or disease or natural disasters.

Weather: Many animal species have gone extinct due to unfavorable weather caused by either humans or natural occurrences. The Bramble Cay Melomys have died out or gone extinct due to climate change. **(5.1)**

Disease: The extinct *Rattus Nativitatis*, also known as the Bulldog rat, used to live on an island called Christmas Island. It was wiped out due to ship jumping rats with protozoan diseases. **(5.2)**

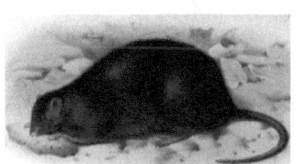

Natural Disaster: The most famous one is extinction due to natural disasters. The famed dinosaurs were killed due to a huge asteroid that hit the earth, wiping the dinosaurs from the face of the earth. The same asteroid may have killed off many more species of which we may not have discovered the remains.

But the question "Can we resurrect extinct animals?" remains unanswered. The answer is "Yes," we can. But that raises more questions than it answers. How can we resurrect them?

There were many attempts in the past at reviving extinct animals. Some attempts have succeeded and some have failed.

Methods of De-extinction: (5.3)

1. Cloning

Dolly, a sheep, is the first cloned animal to survive to adulthood. Many other animals have been cloned, but Dolly the sheep is the most successful cloning.

2. Selective Breeding

Selective breeding is the process by which living relatives of the extinct species are identified and specifically bred to reproduce the traits of the extinct species. This method can recreate the traits of an extinct species, but the new species will differ slightly from the original species.

3. Genome Editing

This method is currently in development. Gene editing tools such as CRISPR-Cas9 can make it possible to edit the genes of a species to make the embryo look exactly like the extinct animal that scientists are trying to recreate.

ASTRONOMY

6

Ten Unique Planets You Won't Believe Exist

Let us learn about ten unique planets outside the solar system you won't believe exist. There are many weird and unique planets in our universe. We have discovered only a few of them. There are many more to discover and learn about.

Number 10: WASP-12b

WASP-12b is the hottest exoplanet—a planet outside the solar system—ever discovered with temperature reaching 2210°C because it is moving in a spiral orbit closer and closer to its star. It is on a suicide spree and it is 870 light-years away from us. A light-year is the distance that light travels in a year which is approximately equal to 9.46 trillion kilometers or 9,460,000,000,000 kilometers.

Number 9: 55 Cancri e

It is an exoplanet largely made up of diamonds. It is twice the size of Earth and eight times as massive as Earth. It is because of what it is composed of.

Number 8: TrEs 2b

TrEs 2b is a gas giant, a large planet composed mostly of gases. It is the darkest planet ever found. Though it is only three million miles away from its host star, it is still very dark. It is darker than a solid piece of coal or the darkest acrylic paint.

Number 7: J1407b

This planet J1407b is also a gas giant. It is 420 light-years away from Earth. It has more than 30 rings. Each of the rings has a diameter of tens of millions of miles.

Number 6: Gliese 436 b

Gliese 436 b is actually a unique planet because it is covered with "flaming ice." It is called the "planet of burning ice." It has clouds made of hydrogen, and it gets burning hot and leaves trails of hydrogen gas behind. This makes it a truly unique wonder.

Number 5: GJ 1214 b

This planet GJ 1214 b is also a unique planet because most of its mass is composed of water ice but it appears to be rather hot. It is six times as massive as Earth.

Number 4: WASP-17b

WASP 17b is unique in the sense that it is the first planet to be discovered that orbits in the opposite direction of its host star. Furthermore, it is the puffiest planet. It is one of the largest planets discovered but it has only the mass equivalent to half the planet Jupiter.

Number 3: Gliese 581 c

This planet named Gliese 581 c made headlines when it was first announced because people thought it was habitable. Then it was later found that it was tidally locked which means one side was always night and the other side always day.

Number 2: HD 106906 b

It is a giant planet and has 11 times the mass of Jupiter and it orbits its star 650 times the distance between Earth and the Sun.

Number 1: HD 188753

This is also unique. HD 188753 is the first known planet to assign a triple star system. This means that it orbits a star which orbits another star which orbits another star. That is really a weird planet. That is why I put it on number 1.

(Source: 6.1)

7
What is the Hottest Object in the Universe?

What is the hottest object in the universe? Is there a limit to heat? If there is, what is the limit to heat? When you think of hot things, you may think of the Sun or something else, but let me introduce a celestial object that is far hotter than the Sun. What if I told you the Sun 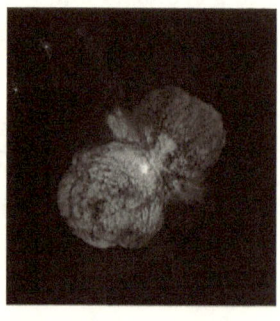 is cold compared to this star? It is actually a star system of two stars. This star system is Eta Carinae, the hottest star system in the known and observable universe. It is also hotter than all other stars. But wait! We can get even hotter.

Let us learn more about stars. Stars can range in temperature from the relatively cool red dwarfs to superhot blue stars that are over 10,000 K. The color of a star is related to its temperature. If a star looks red, it means its surface temperature is approximately 2,500 K. If you look at the Sun from space, it is actually white because it is 6,000 K. The hottest stars are the blue stars. A star appears blue once its surface temperature has reached above 10,000 K. Like absolute zero, there is absolute hot. There are extremes at either end—the coldest and the hottest.

Absolute hot is the hottest temperature possible. And Plank's theory states that when an object gets hotter, its visible wavelength gets shorter. The shortest frequency or wavelength possible is 0.000001 nanometers at which the temperature will be 142,000,000,000,000,000,000,000,000,000,000 K or 142 nonillion K. That is very hot! It is much hotter than the Sun or anything else that our minds can fathom.

When Eta Carinae dies, which might happen anytime, it will send a blast of narrow gamma rays through space that will outshine the entire Milky Way galaxy. And if it happens to hit the Earth, it will destroy ¼ of its ozone layer from 80,000 light-years away.

That is the power and capability of Eta Carinae. We may discover hotter stars and star systems, but for now, Eta Carinae is the hottest star and star system.

(Source: 7.1)

AERONAUTICS

8
How Do Airplanes Fly?

Mankind had been trying to take flight since the 18th century. There were many ridiculous inventions to take flight. For example, bicycles with wings, wings attached to arms, and so on. None of these inventions worked. The first successful attempt was on December 17, 1903, when the Wright Flyer took off from Kitty Hawk, North Carolina.  (8.1) The secret of its success was its airfoil technology. The airfoil of the wing allowed the wing to funnel the air downwards and produced lift which in turn helped the aircraft take off.

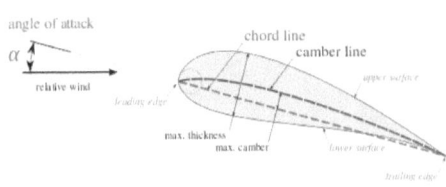

The application of airfoil technology can be seen in modern aircraft as well. The principles behind the airfoil technology are Bernoulli's principle and Newton's third law of motion. The wings generate lift which helps the plane take off. Now, we have seen the basic principles which help the plane take off, but we shall see what propels it through the air. To propel anything forward, you need thrust. Planes produce thrust by sucking the air in from the atmosphere and put it through a compression chamber. Then the compressed air goes through the combustion chamber where it is mixed with fuel, ignited and shot out.

Here you can see the ailerons, flaps, elevators, a rudder and so on. These are called control surfaces. Control surfaces help pilots maneuver the plane. There are names for the movement of the aircraft, such as pitch, yaw and roll. To take off and land, the pilots use elevators and flaps. The ailerons are used to tilt the plane left or right. This motion is called "roll." **(8.2)** The movement of turning the plane left or right using the rudder is called "yaw." But the pilots do not turn the plane like we steer a car because it will disorient the passengers and cause discomfort. The elevators are used to control pitch, that is, the tilting of the nose up or down.

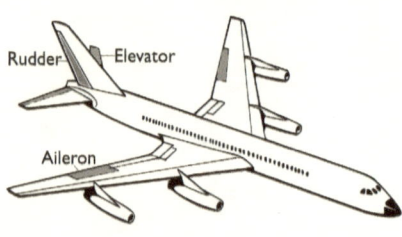

A cockpit is near the front of an airplane from which pilots control an airplane. You can see the throttle which helps increase or reduce speed. In a plane with two engines, there are two throttles, one for each engine.

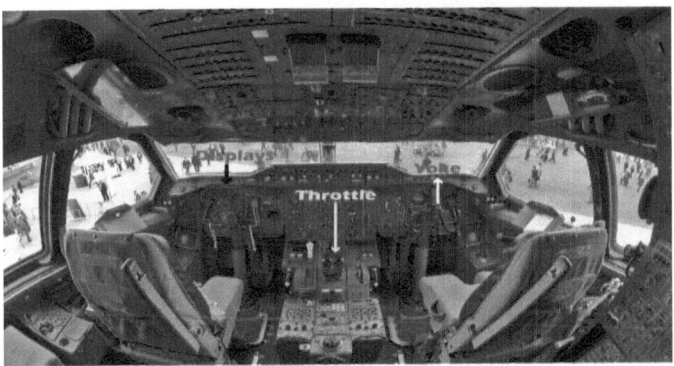

There are displays to show altitude, speed and so on. There is something that looks like a steering wheel of a car, it is known as a "yoke." The yoke helps pilots make the planes take off, land and turn left or right. **(8.3)** And there is a button for airbrake to reduce the speed of the plane while landing. There is also a radio to talk to air control so that the pilots are all synchronized when taking off and landing. This provision also ensures that they are not taking off or landing on the same runways and so on. The displays also display the position of the aircraft so that the pilots know the position of the aircraft to help them maneuver the aircraft.

You may be wondering where the planes store fuel. They store it in the wings in order to have proper weight distribution and balance thus to even out the weight. The wings are perfectly suited for storing fuel. **(8.4)**

Military fighter jets have a lot more factors, such as "wave suppression" and "stealth" that passenger aircraft do not have. But, basically, all airplanes are almost the same.

As God designed birds symmetrically and aerodynamically, humans have built planes. The airplane is an incredibly complex flying machine that we should never take for granted. A century of work has led to the making of a near-perfect modern airplane.

9
How Do Helicopters Fly?

Helicopters are incredibly complex flying machines. They have a huge history behind them. It started way back in 400 B.C. when a Chinese book described a flying top which was basically a stick which had wings attached to it. When it was spun quickly between hands, it flew like a helicopter off the ground. The following attempts are some of the important milestones in the history of helicopters. Leonardo da Vinci (1452–1519), the painter who painted Mona Lisa, was a very influential personality during the Renaissance. He designed a helicopter with corkscrew-shaped propellers to lift it up through the air, but he never succeeded. Then Sir George Cayley built a spring elastic model of a helicopter that flew to tens of meters. Thomas Edison experimented with model helicopters, including some driven by electricity. There was an attempt that led to success. Igor Sikorsky (1889–1972), a Russian-American aircraft engineer, built a working model of a helicopter powered by a rubber band. It was the beginning of a life's obsession

with helicopters. After many years, Sikorsky designed and patented a working helicopter. He finally built and flew a version of the VS-300 (Vought-Sikorsky-300) in 1940, and in 1942 his model became the world's first mass-produced helicopter. **(9.1)**

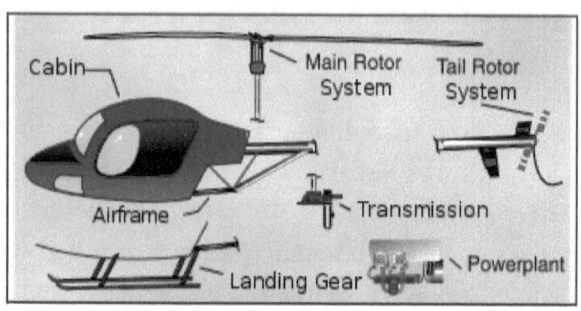

Helicopters are very complex both inside and out. Sikorsky spent a lot of time developing and patenting his model. Let us now look at the main parts of a helicopter and their functions. The main parts of a helicopter are the engine, landing skids, airframe, tail rotor and the main rotor which consists of a mast, hub and blades. **(9.2)**

The main rotor's function is to help propel the helicopter forward and up. The helicopter's blades are tilted in various angles using swashplates. The rotor blades use the airfoil shape and the working of the airfoil technology has already been mentioned in "How Do Airplanes Fly?" A helicopter lifts off the ground using the rotor. Using the swashplates, it tilts its nose forward a little so that the rotor is facing both down and backward providing it thrust to move forward and hover above the ground.

You may have noticed a tail rotor spinning. Let me tell you the use of this tail rotor. We are all familiar with Newton's third law of motion and it states: "For every action there is an equal and opposite reaction." As the rotor spins in one direction, the helicopter spins in the opposite direction, and to solve this issue there is the tail rotor.

The tail rotor is located at the very end of the tail to have more effect with less thrust. It controls the spin of the helicopter so that the helicopter does not spin uncontrollably. Pilots use the

spin of the helicopter and the tail rotor to their advantage. They turn the throttle of the tail rotor down so that it does not have the counterbalance effect making the helicopter turn to one side. If the pilot turns the throttle up a lot, it turns to the other side. This motion is called "yaw." We have seen the working of the helicopter.

Now, let us see the inside of the cockpit. You can see the foot pedals which are known as anti-torque pedals. They are used to control

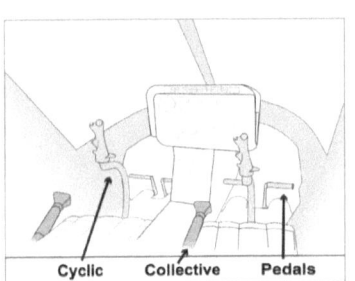

the yaw motion. There are things that look like car gear sticks. Those are the cyclic pitch and collective pitch levers which are used to control and maneuver the helicopter. There is a throttle, as in airplanes, to control the speed of the rotation of the rotors. There are buttons and switches for lights and the like.

The following are some differences between helicopters and airplanes:

- Helicopters do not need runways. So, they can take off and land in a small amount of space. Helicopters can even land on big yachts.
- Helicopters' main thrust is upwards whereas the planes' thrust is forward.
- As planes have wings, helicopters have rotors.
- Piloting an airplane is very difficult, but piloting a helicopter is relatively harder.

The invention of helicopter shows how far humans have advanced in science and technology. It is a modern marvel and its power cannot be underestimated.

10
Mankind's Quest for Speed

Mankind has had a never-ending quest to be better. Ever since the beginning of the human race, humans have been trying to find ways to be more efficient. Likewise, humans have also been trying to be faster. The pursuit of speed seems to be in our blood.

Humans first started domesticating animals around 10,000 years ago. The first animal to be domesticated was the goat. At first, humans domesticated animals for meat and eggs. Thousands of years later, humans started domesticating animals for transportation. The first animal to be domesticated for transportation was the horse and it happened around 4000–6000 years ago. We domesticated horses for convenience, speed and efficiency. Then we invented cycles in 1817. Cycles were preferred by working-class men and women in the late 1800s.

The most important advancement in the history of speed is the introduction of cars. The first car, the Benz Patent Motor Car, was introduced in 1886. Karl Benz

invented it. Karl Benz is the founder of Mercedes Benz. The Benz Patent Motor Car had a top speed of 16 kilometers per hour (km/h) or 10 miles per hour. Technology in the automobile industry advanced rapidly in the early 1900s through the 1930s. The fastest car achieved a speed of 16 km/h in 1886, 80 km/h in 1902, 192 km/h in 1928 and it jumped to a huge speed of 432.7 km/h in 1938. The record of 432.7 km/h on a public road was never beaten until 2017 when an Agera RS beat the record.

The land speed record is a highly contested record. Most of the land speed records were set or broken in the Bonneville Salt Flats. The Bonneville Salt Flats are a vast expanse of densely packed salt pan located in Utah in the USA. It was first used for setting speed records in 1912. The fastest land speed record to this day was set by Andy Green from the USA. He set his record in the Thrust Supersonic Car (SSC). The Thrust SSC reached a whopping speed of 1,227.986 km/h. But, this speed record was not

set on the Bonneville Salt Flats. The fastest record on the Bonneville Salt Flats was set by Gary Gabelich in the rocket-powered Blue Flame. The Blue Flame reached a top speed of 1,014.656 km/h. Gary Gabelich was the first person to break the speed of 1000 km/h.

Safety has been a big concern in commercial cars. In the early automobile era, seat belts were not mandatory in cars. Seat belts were an additional feature. Then people realized the importance of seat belts, which were then made mandatory. To improve safety in cars, car manufacturers run crash tests. Nowadays, crash dummies are used 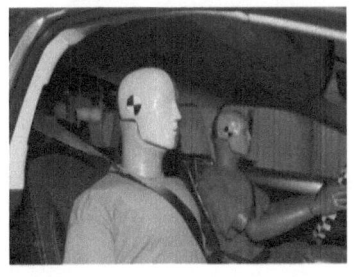 to replace humans. Back when crash dummies were not perfected, scientists and engineers would replace the crash test dummies and put themselves in dangerous situations. Not long afterward, many advances were made in the field of car safety, such as the provision of airbags and ABS (Anti-Lock Braking System). Airbags are deployed when high forces are detected. The aim is to protect the head of a driver from crashing into the steering wheel of a car. The ABS is a braking mechanism that helps save many lives. Actually, other braking systems stop the wheel fully and thus stop the car. The ABS alternates between braking and releasing so that the car slows down but does not stop completely. This gives the driver the ability to steer away from the crash. Mankind has come a long way from domesticating horses for transportation. We can only imagine what great advances may come in speed and safety in the near future.

(Source: 10.1)

11
Automobile Industry

The growth of the automobile industry is immense worldwide. The activities of the industry mainly include design, manufacturing and marketing. Here is a brief description of the brands of the five most popular car companies: Ferrari, Lamborghini, Bugatti, Tesla and Pagani. I am mentioning only those car companies that have made great innovations in making high-end cars.

The legendary Italian car company Ferrari manufactures fast cars with an engine capacity of more than 300 hp (horsepower). They have made different models over the past, such as Ferrari La Ferrari, and Ferrari F40. Ferrari F40 has an attractive, sleek and aerodynamic design as well as great speed.

Lamborghini is a car company based in Italy. They make expensive high-performance supercars. They keep their library of cars limited. Currently, they manufacture only three types of cars: the Huracan, Aventador and Urus. Let us start with the Huracan. The Huracan is one of the few cars Lamborghini manufactures. It features a

5.2-liter V10 engine with a fuel capacity of 80-83 liters, and it is one of the best supercars in the world. Up next, the Aventador is one of the iconic Lamborghini cars ever made. It features a 6.5-liter V12 engine with a fuel tank capacity of 90 liters. Lamborghini  Aventador is one of the fastest cars in the world. Next is the Urus. The Urus is an SUV (Sports Utility Vehicle). It features a 4.0-liter V8 engine and has a fuel tank capacity of 85 liters. **(11.1)** Let us take a look at the history of Lamborghini. It was founded in 1963 in Sant'Agata Bolognese, Italy, by Ferruccio Lamborghini. It is owned by Volkswagen through Audi. Lamborghini also owns the bike company Ducati. Lamborghini started out making tractors. Then it transitioned into making cars that competed with the titans of the car world such as Ferrari and the like. They found great success in the Miura Sports Coupe which shot them to fame. Currently, Lamborghini produces sports cars and V12 engines for offshore powerboat racing. **(11.2)**

List of Past Lamborghini Cars

Lamborghini Miura (1966-1972), Lamborghini Islero (1968-1969), Lamborghini Urraco (1973-1979), Lamborghini Espada (1968-1978), Lamborghini Countach (1990-1974), Lamborghini Jalpa (1981-1988), Lamborghini Diablo (1990-2001), Lamborghini Murcielago (2002-2010) and Lamborghini Gallardo (2003-2013). **(11.2)**

Now let me tell you about the French car conglomerate Bugatti. They make very fast and sleek cars. The Bugatti Veyron is one of the fastest cars in the world.

The fastest motorcycle in the world is Dodge Tomahawk. If the Bugatti engineers keep their hands on anything, it goes faster.

Tesla, owned by tech legend Elon Musk, is an electric car company. Tesla is revolutionizing the automobile industry. They have made their cars more beautiful than ever and they have hit the ground hard with Tesla Roadster 2020. They are going to release it in 2020 for which you can have a test drive right now.

Pagani Zonda is a sports car manufactured by the Italian car company Pagani. The Pagani Zonda is very loud. Is it loud for no reason? It is noisy due to the very powerful engine that powers the Pagani Zonda. It has acceleration like that of the Roadster 2020. The Tesla Roadster 2020 is the fastest accelerating production car ever made.

Now we have seen the best of the car world.

ATOMIC SCIENCE

12

Nuclear Energy

As the demand for electricity increases rapidly, we need to explore new and innovative ways to produce electricity. The sources of electricity are solar energy, wind energy, energy obtained from burning fossil fuels and energy released during nuclear fission or fusion (nuclear energy).

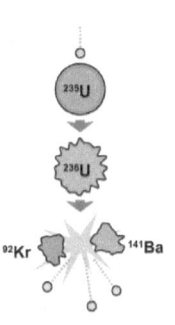

Nuclear energy is the energy released by a controlled chain reaction of fission or fusion of uranium atoms in a nuclear reactor. It releases energy and an enormous amount of heat. Then a coolant, usually purified water, is sent to cool the nuclear reactor core. The heated water produces steam which is used to turn turbines. Consequently, electricity is produced. Ever since the invention of nuclear weapons, nuclear energy has been a hot topic.

The biggest concern with nuclear energy is nuclear waste disposal. The "half-life" of uranium is around

25,000–700,000,000 years. **(12.1)** The "half-life" refers to the period in which the radioactive material stays radioactive. Nuclear reactors are expensive to build. The installation of nuclear plants often faces public opposition because of the role of nuclear energy in nuclear weaponry and nuclear waste disposal.

Nuclear energy has the potential to be the number one source of energy. Nuclear energy can produce way more energy per ton than energy produced from coal or oil. By initiating new researches in the field, we will be able to harness nuclear power as a good source of energy. The use of thorium rather than uranium could reduce

waste and produce more power than uranium. Such nuclear inventions can revolutionize the way we produce electricity. Nuclear energy is not worth it unless we can make it safe.

Advantages and Disadvantages of Nuclear Energy

Advantages:

Low Greenhouse Gas Emission

In terms of carbon emissions, nuclear energy produces less greenhouse gases than energy produced from coal or oil. The greenhouse gases are chlorofluorocarbons (CFCs), hydrofluorocarbons (HFCs), ozone, methane and carbon dioxide. The greenhouse gas emissions have reduced a lot because of the prevalence of nuclear power. Thousands of lives have been saved as coal power plants have been replaced with nuclear power plants.

Reliability

It is estimated that the amount of uranium left in the earth will last for 70–80 years. A nuclear power plant can run uninterrupted for over a year. **(12.2)** Renewable forms of energy, such as solar and wind energy, are dependent on weather conditions. But, nuclear power plants can run independently without climate disruption.

Some Other Advantages:

- Surprisingly, deaths due to nuclear power, even taking nuclear accidents into account, are lower compared to the deaths caused by power generated from coal, oil and other methods.
- With new advances in nuclear technology, we will be able to produce pollution-free, efficient energy.
- Once we perfect methods of producing nuclear energy, we will not likely have an energy crisis later.

Disadvantages:

Environmental Impact

One of the biggest problems is the environmental impact in relation to the mining and transportation of uranium. Mining, refining and transporting uranium are expensive and dangerous. Uranium has to be transported in special vehicles with reinforced barriers to prevent leaks. **(12.2)**

Nuclear Waste Disposal

Nuclear waste disposal poses a great danger and it is hard to do. You cannot throw nuclear waste which is highly radioactive to a landfill. Due to its radiation, it will kill every living thing around it. To resolve this problem, many solutions, such as putting nuclear waste deep underground and the like have been suggested, but nothing has come to fruition. Currently, most nuclear power plants store their waste locally. Over a long period, nuclear waste decays to low radioactivity. Reprocessing of nuclear waste can be done, but nuclear power plants cannot make power out of that. The reprocessed material must be heavily guarded as it can be used to make nuclear weapons.

High Cost

At present, nuclear energy is costly and difficult to produce. Nuclear power plants are also very expensive to build. This reason and public opposition combined may have reduced nuclear power funding from 18% in 1996 to 11% in 2015.

Non-renewable Nature of Uranium

Like coal and oil, uranium is a non-renewable energy resource. Uranium exists in only a few countries. It is also expensive to mine and process.

Some Other Disadvantages:

- Exposure to radioactive materials can cause cancer and a plethora of other diseases.
- Nuclear accidents happen every 15–20 years. Nuclear accidents such as Chernobyl and Fukushima Daiichi are devastating, as they have claimed the lives of many workers. Crops are unable to grow and life will not be able to thrive in places of nuclear accidents for thousands of years.
- Unless nuclear technology is perfect, widespread use of nuclear technology is unlikely.
- Nuclear plants are targets of military attacks.
- Nuclear power plants can be put to use for making nuclear weapons.

Conclusion

Nuclear technology will inevitably get better in the near future. When nuclear technology is improved, nuclear energy may be the top source of energy in the world.

SPACE SCIENCE

13
The Achievements of NASA

NASA stands for National Aeronautical and Space Administration. NASA is the space travel organization of the USA. Other countries have space travel organizations of their own. Japan has JAXA (Japan Aerospace Exploration Agency) and India has ISRO (Indian Space Research Organization). NASA launched its first satellite in 1958.

Here are the top eight achievements of NASA:

8. Explorer 1

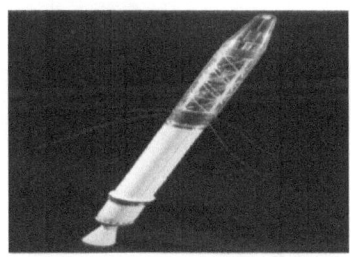

NASA's first satellite was called Explorer 1. Explorer 1 was launched from Cape Canaveral Air Force Station. It was equipped with a cosmic-ray detector and a micrometeorite detector. **(13.1)** It was launched as an interesting sequel to the USSR's (Soviet Union) Sputnik. Explorer 1 was launched due to a space race between the USSR and the USA. The USSR was the first to reach space and put a human in an orbit. The USA was the first to put men on the moon.

7. Chandra X-ray Observatory

The Chandra X-ray Observatory is a part of NASA's fleet of space telescopes. It was launched in 1999. It was named after the  revolutionary Indian-American scientist Subrahmanyan Chandrasekhar.

The Chandra Observatory is to X-ray astronomy as Hubble is to optical astronomy. The Chandra X-ray Observatory has provided us with many discoveries. It focuses X-rays by using four pairs of nested iridium mirrors. It can send beautiful high-resolution images back to Earth. Chandra's primary focuses are black holes and supernovas. Chandra is managed by the Chandra X-ray Observatory Center. **(13.2)**

6. Hubble Space Telescope

The Hubble Space Telescope was launched in 1990. It is undoubtedly the most iconic telescope of all time. It gets its power from the sun using solar panels.
Steadiness is the key to take photos of or observe planets, distant galaxies or stars. It is stated in one of NASA's articles that "the telescope is able to lock onto a target without deviating more than 7/1000th of an arcsecond,

or about the width of a human hair seen at a distance of a mile." **(13.3)** The Hubble telescope sends 10 terabytes (TB) or 10,000 gigabytes (GB) of data every year. Astronomers have published 12,800 scientific papers using data gathered from the Hubble Space Telescope. It is about the size of a school bus. It does not require thrust to move. **(13.3)**

5. Pioneer 10

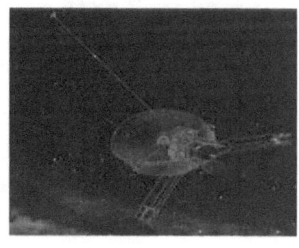

Pioneer 10 was a mission to reach Jupiter. It was launched in 1972. It provided crucial information about Jupiter. It worked from 1972 to 2003. Communication was lost in 2003 due to the loss of electric power. Following its encounter with Jupiter, Pioneer 10 explored the outer regions of the Solar System, studying energetic particles from the Sun and cosmic rays entering our portion of the Milky Way.

4. Apollo 13

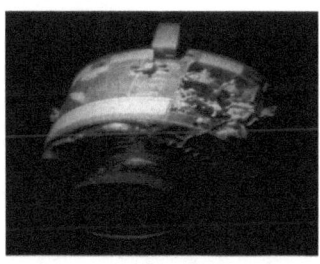

Apollo 13 was a mission intended to land on the moon. The plans were aborted after an oxygen tank exploded and damaged the Service Module. The Command Module, which was an essential part of the spacecraft, depended on the Service Module.

The accident occurred about 56 hours into the mission when the live TV broadcast was about to close. But, the spacecraft successfully entered the Earth's atmosphere and splashed down in the ocean. NASA learned from its mistakes and an extensive revision of the oxygen tank which exploded was ordered. NASA took many corrective actions to make sure such a thing never happens again. **(13.4)**

3. Freedom 7 or Mercury-Redstone 3

It was the USA's first manned mission to space. It was piloted by Alan Shepard. Its goal was to put a man in an orbit and return to Earth safely. It was launched from Cape Canaveral LC-5. The mission only lasted 15 minutes. Freedom 7's launch was watched by an estimated

number of 45 million people. It was a great achievement for its time. Those 15 minutes was an iconic time in NASA's history. **(13.5)**

2. International Space Station (ISS)

The International Space Station is a space station or a habitable artificial satellite in low Earth orbit. It was built piece by piece. Its first component was

launched in 1998. It was called the ISS Zarya. The ISS is basically a means for astronauts to live in space. The ISS is the largest structure outside the planet Earth. The ISS has a camera attached to its bottom which livestreams from space to Earth. It can be viewed on livestreams called ISS Livestreams.

1. Apollo 11

Apollo 11 was the mission that put humans on the Moon. It was launched in 1969. It was the biggest undertaking of NASA during that time. NASA had tirelessly worked on it. It used a rendezvous to land humans on the Moon. Astronaut Neil Armstrong was the first man  to step on the Moon. On the Moon, he uttered the words that would echo in our ears for years: "One small step for man, a giant leap for mankind."

14
NASA's Plan to Return to the Moon and Stay

NASA has just recently announced its plan to return to the Moon, 50 years later. The plan was announced on February 14. The plan will require an additional $1.6 billion in order to make it come to fruition. This plan will start in 2024 and end in 2028.

The spacecraft will put the first woman on the Moon. The capsule, which will carry the humans, will be called "Orion." The mission as a whole will be called "Artemis." Artemis is the twin sister of Apollo and also the goddess of the Moon in Greek mythology. Apollo 11 was the first successful mission to put man on the Moon. NASA plans to use an orbital station in order to get the lunar lander to the Moon. The orbital station will be named "Gateway." NASA's administrator Jim Bridenstine said, "We're doing it in a way we've never never done before." NASA wants humans to return to the Moon on the 50th anniversary of the Apollo 11 Moon Landing. **(14.1)**

NASA is relying on commercial launch providers or private spaceflight companies, such as SpaceX and Northrop Grumman Innovation Systems, in order to bring this plan to fruition. **(14.2)** NASA also says that they may start launching payloads as early as this year. NASA has released an official YouTube video announcing the plan. To watch this video, go to this shortened link: bit.ly/ArtemisMission. **(14.3)**

I believe that this will propel humanity into a new age, a new era of space exploration. To think that this could be a reality in 5–10 years is amazing. This is an example of accelerating returns. Accelerating returns is a concept of very fast advancement. To put it in context, we have achieved 30 years of technological advancement in 15 years. At this rate, in the next 100 years, we will multiply our technological power by 20,000. Let us see the differences between the original Moon Landing (Apollo 11) and Artemis. The Apollo 11 mission used a rendezvous method in order to land on the Moon. Artemis will use an orbital station in order to get to the Moon. The Apollo 11 was the result of a space race between the USSR and the USA, but the Artemis is an international collaboration. I believe that Artemis marks a new chapter in the history of spaceflight.

THERMODYNAMICS

15
Absolute Zero

Absolute zero is the idea of the coldest temperature on the Kelvin Scale. William Thomson, the first Baron Kelvin, confirmed the theory that heat is molecular movement. When molecular movement slows, an object becomes cooler. In absolute zero, the particles in an object will have no movement. It is impossible to attain absolute zero because of the Heisenberg Uncertainty Principle which states that we cannot know both a particle's position and its momentum precisely at the same time. If we precisely know a particle's momentum, we cannot know its speed, and vice versa. If we attained absolute zero, we would know both the momentum and position which would both be zero. We have not attained absolute zero, but have we got close to it? We have got 1/1,000,000,000th (one-billionth) of a Kelvin close to absolute zero in a lab. You may be wondering what the coldest natural place in the universe is. It is the Boomerang Nebula. The Boomerang Nebula has a temperature of 1 K. The microwave radiation

from the Big Bang spread evenly. As a result, space has an average temperature of 2.7 K. The Boomerang Nebula spews out gases and is exposed to minimal microwave radiation. That is why it became the coldest place in the universe. Now you know what the coldest place in the universe is. Though we have trillions and trillions of kilometers to explore, we are sure that the Boomerang Nebula is the coldest place in the universe.

Conclusion

- Absolute zero is the lowest theoretically possible temperature.
- Absolute zero is a state where no energy (heat energy) is produced.
- Absolute zero is not physically obtainable.
- Absolute zero means no molecular or atomic movement.
- The coldest obtainable temperature is 1 K.

(Source: 15.1)

NEUROSCIENCE

16

How Does Our Brain Store Memories?

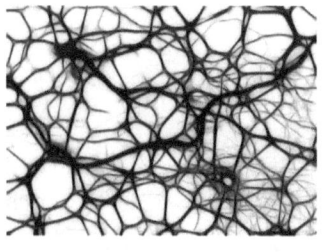

The brain is an incredibly complex organ and it comprises many parts. There are billions of neurons inside the brain. Neurons store information. Understanding the function of neurons is essential to understanding how the brain stores memories. Memories are a set of neurons that are activated when an activity is done. We know what memories are. But how does our brain store them? When an activity is done, a specific set of neurons is activated. When these neurons are activated again, you can remember the activity associated with them. As the same neurons are activated repeatedly, they get closer to one another and the memory is strengthened.

Let us see where our memories are stored.

There are two types of memory: long-term memory and short-term memory.

Long-term Memory

Monumental events that happened in our life, such as the first time we rode a cycle and the first day at school are stored in the hippocampus, so that they last for a long time.

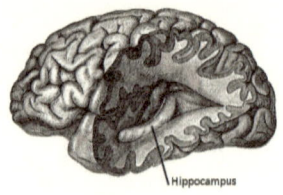

Short-term Memory

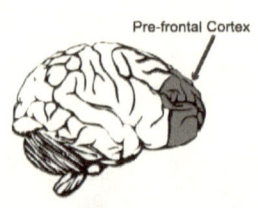

The short-term memories are stored in the pre-frontal cortex. For instance, insignificant and repetitive activities, such as drinking a soft drink and drinking water.

How much information can be stored in our brains?

There are over 100 billion neurons in our brain and they have the potential to store almost an infinite amount of memories.

The brain has such intricate and incredible complex ways of storing memories. I have given here only the gist of it. Some people say that they have bad or inferior memory, but that is not true. Everybody has an amazing brain inside them. Everybody has a supercomputer in their heads. If we use our amazing brain properly, there is no limit to human advancement.

The brain is very intricately designed. This is a proof of an intelligent designer—God.

(Source: 16.1)

MODERN PHYSICS

17
Higher Dimensions

What is meant by higher dimensions? They are dimensions beyond our perception and comprehension. These dimensions are theoretically possible.

Let us start with dimensions that we can perceive.

First, let us see two dimensions. What we view on our cell phones and computers is a two-dimensional image of a three-dimensional world. That is, a three-dimensional object drawn on paper becomes two-dimensional, for example, a prism, a sphere and a cone. Now, let us imagine a flat world as described in Edwin A. Abbott's book *Flatland: A Romance of Many Dimensions.* In his book, Abbott describes a flat world with geometrical shapes as its inhabitants. They cannot perceive depth, the third dimension. To them, light is a mystery as it comes from above which requires a third dimension. If you could put your hand through the two-dimensional world, it will look as if your hand is shape-shifting. It is the same with four-dimensional objects. When we see a four-dimensional object, it will look as if it is shapeshifting. **(17.1)** Take a look at

this four-dimensional cube also known as tesseract or hypercube. If you want to see the animated GIF of the rotating tesseract, visit this link: https://en.wikipedia.org/wiki/File:Tesseract.gif#/media/File:Tesseract.gif

There is an interesting theory about certain things called wormholes. Wormholes are things that warp spacetime. If we imagine space like a piece of fabric as Albert Einstein states in the theory of relativity, wormholes will be possible. Wormholes bend spacetime so that we can take a shortcut through space. It is almost impossible to keep a wormhole open as gravity tries to close it. We need something called "exotic matter" to keep the wormholes open. Exotic matter is not antimatter. The exotic matter has negative mass, unlike normal matter which has positive mass. **(17.2)**

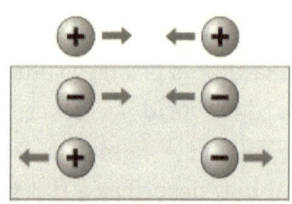

The atoms in exotic matter repel, that is, they do not attract. This could keep wormholes open. Wormholes may make it possible to traverse dimensions. The assumption that wormholes exist is just a theory. It exists on paper and it may be virtually impossible in real life.

Now, we have traversed the nooks and corners of what math deems possible. Is it math? Is it true? We may never know.

18

Antimatter

Antimatter is matter but the opposite of it. Matter is made up of protons and electrons. In matter, the proton is positively charged and the electron is negatively charged. In antimatter, there is an antiproton and there is a positron. The antiproton is negatively charged and the positron is positively charged. Antimatter is incredibly expensive to make and contain. NASA puts the estimate of the price of 1 gram of antihydrogen at $62.5 trillion (adjusted for inflation $94 trillion). This makes antihydrogen (the opposite form of hydrogen) the

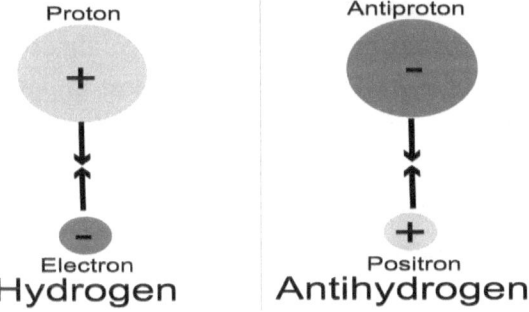

most expensive material in the world. It is highly costly due to the very difficult process of making antimatter. Antimatter may seem amazing, but it gives rise to the question "Does it have any actual uses?" PET (positron emission tomography) uses positrons to produce high-resolution images of the body. It can detect and even kill tumors.

Ever since the screening of the famous TV series "Star Trek," people have been fantasizing about antimatter propulsion. Using an equal amount of matter and antimatter and using annihilation (the process of colliding matter and antimatter resulting in the transfer of 100% of their mass into energy) planes and rockets can be propelled. Antimatter may have more uses we are yet to discover. Antimatter has a lot of potential.

Using antimatter propulsion, we can possibly achieve interstellar or intergalactic travel. Antimatter has the potential to launch our civilization into a new age, the age of intergalactic travel. But before all this, we must find proper ways to make antimatter and harvest its power. As of now, making a few atoms of antimatter and storing them is very difficult. If we can produce antimatter at a relatively low cost, there is no limit to our advancement. Using antimatter propulsion, we can go almost 50% of the speed of light. Currently, it is almost impossible to produce even a gram of antimatter. In the future, we may be able to make and store antimatter with relative ease, but until then we can keep trying.

(Source: 18.1)

19
The Story of a Genius: Albert Einstein

Albert Einstein was born in Ulm, Germany, on March 14, 1879. He was born to Hermann Einstein, a salesman and an engineer, and Pauline Koch. Einstein excelled in Physics and Maths at a very young age. He mastered integral and differential calculus by the age of 14. **(19.1)** The year 1905  was monumental for Albert Einstein. It is called Einstein's Miracle Year or Annus Mirabilis.

It was the year that the four most important papers of Einstein were published. Those four papers were "On a Heuristic Viewpoint Concerning the Production and Transformation of Light," "On the Motion of Small Particles Suspended in a Stationary Liquid, as Required by the Molecular Kinetic Theory of Heat," "On the Electrodynamics of Moving Bodies" and

"Does the Inertia of a Body Depend upon Its Energy Content?" These four papers laid the foundation for modern physics and were a cornerstone of quantum mechanics.

The first paper "On a Heuristic Viewpoint Concerning the Production and Transformation of Light" was on the photoelectric effect for which he received the Nobel Prize in 1922. The second paper "On the Motion of Small Particles Suspended in a Stationary Liquid, as Required by the Molecular Kinetic Theory of Heat" was about Brownian motion proving atoms' existence. The third paper "On the Electrodynamics of Moving Bodies," was about special relativity which was one of the greatest achievements of his life. The fourth paper "Does the Inertia of a Body Depend upon Its Energy Content?" was about the mass-energy equivalence. In his fourth paper, Einstein wrote probably the greatest formula of all time, $E=mc^2$. **(19.2)**

These four papers are the most important ones, though Einstein wrote over 300 scientific papers and 150 non-scientific papers in his lifetime. In 1933, Einstein visited the USA which was also the year Adolf Hitler came to power. Einstein, being an Ashkenazi Jew, did not return to Germany. He settled in the USA and became its citizen in 1940. He was one of the leading causes behind the "Manhattan Project" which led to the development of nuclear bombs. Einstein's formula $E=mc^2$ was the main principle behind the making of the atomic bomb.

The first atomic bomb was dropped on Hiroshima and the second one on Nagasaki, the two Japanese towns, in 1945 which left lakhs of people killed and injured. This incident exacted a heavy price on Einstein, immersing himself in depression, as he was one of the causes  of making the bomb. Einstein is one of the greatest physicists who have ever lived on this planet Earth. Without him, the world would not have seen many modern inventions.

ENVIRONMENTAL SCIENCE

20
Global Warming

Global warming is a rise in the Earth's average temperature. There are many impacts of global warming. Some examples are rise in sea levels, increased rate of wildlife extinction and extreme weather conditions. The rise in the sea level is caused due to the polar ice caps melting. When the sea level rises, places close to the ocean will get submerged. Some islands in the Pacific like Tuvalu may be almost completely submerged in water due to the rise in sea levels. The wildlife living in the Arctic and Antarctic may be wiped out due to the destruction of their habitat. The destruction of their habitat (ice) is due to the polar ice caps melting. Since 1906, the global average temperature has risen by 0.9°C. Due to climate change, we may face colder winters and hotter summers. We now know the effects of global warming.   Let us see how it happens and how we can stop it. To understand how global warming happens, first, we have to know what greenhouse gases are. Greenhouse gases are gases in the atmosphere which trap the heat of the sun so that the Earth does not freeze during the night. Greenhouse gases are very important to us. Even water vapor is a greenhouse gas. When there are more greenhouse gases, the atmosphere traps too much sunlight and makes the Earth hotter. The smoke from car exhausts and factories and the burning of plastics and rubber burn a hole in the ozone layer and add more greenhouse gases to the atmosphere. In addition, deforestation and fossil fuels also contribute to global warming. We have seen how global warming affects humans and the life around us.

Now let us look at the solutions to global warming.

1. Renewable Energy

The use of renewable energy can reduce pollution and emission of greenhouse gases.

2. Energy Efficiency

If we reduce our consumption of energy, we do not have to produce a lot of energy. And it leads to the burning of less fossil fuels. As a result, it reduces the effects of global warming.

3. Recycling

By recycling used items and by using recycled items, we can reduce our energy consumption. If we recycle more and consume less, we do not need to burn fossil fuels which can reverse global warming.

We have just learned how global warming works, how it affects us and how we can reduce its effects and possibly reverse its effects. Let us all reduce waste, use clean energy and reuse and recycle the products that we use. Let us be more responsible and sensitive to the impacts of global warming.

HEALTH SCIENCE

21
The World Health Organization (WHO)

The WHO is a branch of the United Nations which is concerned with international public health. It was established in 1948. The WHO's headquarters is located in Geneva, Switzerland. The WHO strives hard to meet the United Nations' noble eight Millennium Development Goals. They are as follows:

1. Eradicating extreme poverty and hunger
2. Achieving universal primary education
3. Promoting gender equality and women's empowerment
4. Reducing child mortality
5. Improving maternal health
6. Combating HIV/AIDS, malaria, and other diseases
7. Ensuring environmental sustainability and
8. Developing a global partnership for development

One of the greatest achievements of the WHO is the eradication of smallpox. The WHO launched a huge campaign to eradicate smallpox and it succeeded in its efforts. The WHO's current Director-General is Tedros Adhanom who assumed office on June 1, 2017. Another formidable disease eradication campaign was the polio eradication campaign. According to the WHO, polio eradication is very possible because polio affects only humans; moreover, there is an effective and inexpensive vaccine available. Immunity against polio is lifelong, and the virus can only survive for a very short time in the environment. Polio is very rare in most parts of the world, or more specifically, in developed countries; but in developing countries, polio is a big issue. A few people who suffer from polio are in Afghanistan, Nigeria and Pakistan, which are developing countries. The global effort to eradicate polio is the largest public-private partnership for public health, according to the WHO. In order to eradicate polio, every child in the world must be vaccinated with the polio vaccine. Polio is a very horrible disease, and the WHO is very close to eradicating it. Research and scientific development coupled with concerted action can definitely eradicate polio.

The WHO has other initiatives and programs apart from the polio eradication program. Aside from the WHO partners, that is, the donors around the globe who financially support its initiatives, the WHO partners with many governmental and non-governmental bodies to respond to global health challenges and bring better health and medication to the people of all nations.

The WHO has two main disease eradication campaigns apart from the polio eradication campaign:

the Stop Tuberculosis Partnership, and the Measles and Rubella Initiative.

A great innovation in the field of health science brought about by the WHO is the first malaria vaccine, recently launched in Malawi. The whole African continent struggles with malaria. Malaria is one of the biggest killers on the African continent. The malaria vaccine has now been launched only in Malawi but it is also expected to be launched soon in Ghana and Kenya. According to statistics, malaria claims the life of one child every two minutes. In the African continent, 250,000 children die of Malaria every year. The WHO recommended bed nets and other measures to control or prevent malaria in the last decade, but progress has reversed or stagnated in most parts of Africa. This vaccine is an important innovation, and it helps save the lives of many children and other people in Africa.

The WHO organizes many events in order to spread awareness of fighting deadly diseases and maintaining good personal hygiene. The WHO celebrates many days in order to raise awareness of health and sanitary. They are World Health Day, World Immunization Week and World Malaria Day. Let us strive hard to see a world without contagious, infectious and deadly diseases.

(Source: 21.1)

22
Fast Food or Healthy Food – What Should We Eat?

Food plays a vital role in determining one's health as food is the building block of every cell in the body. Food is also responsible for the proper functioning of the body.

Does it mean that all kinds of food are good for health? Certainly not. It is the healthy diet that determines one's healthy life.

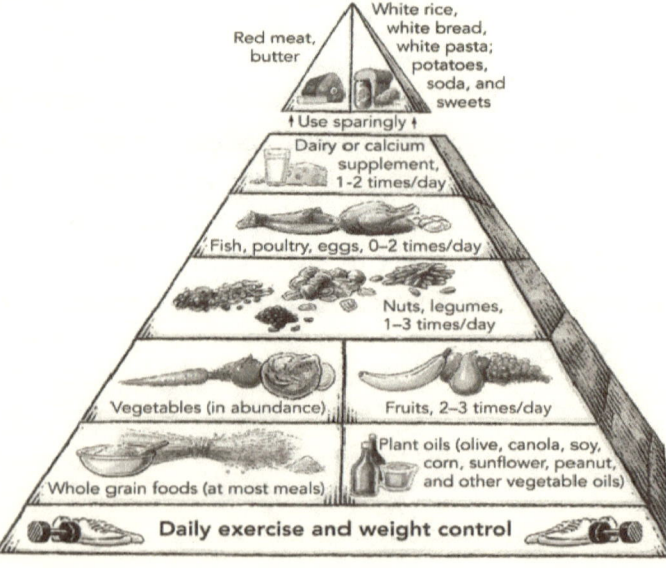

A healthy diet should contain macronutrients (proteins, fat and carbohydrates) and micronutrients (vitamins and minerals) in the right proportion. So what should we eat? We should eat plenty of fruits, green leafy vegetables, whole grains, fish and nuts. We should cut down starchy and sugary food and food rich in saturated fat, such as red meat, butter and cheese. We should drink plenty of water daily. We should completely avoid foods containing chemicals, preservatives, artificial colors and artificial flavors.

Fast food and junk food are very popular these days. Children, and even adults, are fond of eating fast food and junk food. Have you ever looked at the nutritional contents in fast food and junk food? Fast food and junk food are high in calories and sodium and contain many saturated fats. Fast food and junk food cause obesity, diabetes, heart disease and various health disorders. If we eat fast food, we will go out of the world fast! So to live a long and healthy life, fast food and junk food should be completely avoided.

Not all that glitters is gold. Likewise, not all that is tasty is good. The incident that happened in the Garden of Eden can be cited here as an apt example. The forbidden fruit in the Garden of Eden looked very attractive and mouth-watering, and it was pleasing to the eyes of Eve. So, she felt that it was good to eat. Alas! She did not realize that the beautiful fruit was forbidden by God as it contained evil in it. Likewise, attractive and tasty fast food is the forbidden food. It is not the taste of the food that matters. It is the healthy nutrients that the food contains matter.

Good health is the real jewel of life, the most precious possession of man. Therefore, we should take care of our health to live a long and healthy life.

So let us eat plenty of fruits, green leafy vegetables, whole grains, fish and nuts and drink plenty of water and say "No" to fast food and junk food.

23

Does Advertising Influence the Food Choices of Children?

Does advertising influence the food choices of children? I would like to answer with an emphatic "Yes." Yes, verily, advertising influences the food choices of children.

The instruments of mass media, such as television, print and the internet, are wonderful sources of information, but they also influence our decisions and choices.

The images that we see and the messages that we hear have a profound impact on the choices that we make, and specifically the choices that we make with respect to food. And children are mostly influenced by the advertisements that appear on television and in newspapers.

Food manufacturers spend lakhs of rupees on advertisements that contain enticing and catchy phrases and dialogues with appearances of celebrities and models. Advertisements create a powerful impression on children. So the temptation to buy foodstuffs like pizza, burger and hot dogs and soft drinks like Pepsi and Coca Cola is irresistible. Children, and even adults, are not wise enough to make a fair judgment with respect to the choice of food. Many times, children are carried away by the whims and fancies of the advertisements, unmindful of the presence or the absence of the healthy nutrients that the food items contain.

I hope you must have heard of the 1972 KFC advertisement: "Grab a bucket of chicken and have a barrel of fun." **(23.1)** But in reality, you will not have a barrel of fun, instead, you will only have a barrel of flesh in your body.

Most of the food advertisements do not present the true picture of the products. As the information relating to the product is inaccurate and untruthful, children become victims of deceptive and false advertisements.

My advice to every child who reads this is, on any account, please, do not sacrifice your health at the altar of false and deceptive advertisements.

My wish is that every one of us must have the discerning power to make the right eating choices without falling prey to deceptive advertisements.

COMPUTER SCIENCE

24

Parts of a Computer and Their Functions

Let us look at the parts of a computer and their functions. Whether it is a laptop or minicomputer or all-in-one computer, they have many parts in common. Understanding these parts will help you make an informed purchase, that is, make a good decision on buying PCs (personal computers), and also it will help you upgrade or repair your PCs.

First, let me explain the computer case, also known as housing, which houses the motherboard, graphics card and CPU (central processing unit) in a desktop PC. The housing does not include monitor, keyboard and mouse. But in a laptop, the  keyboard, mouse and monitor are in the housing. Inside the housing, all the processing happens and the result or the visual representation of the process is displayed on the monitor.

Next, let us learn more about the motherboard. The motherboard is a part of a computer that wires all the parts together. It is a circuit board to which the CPU and RAM (random access memory) are attached. Every component of a computer, except the case, is connected to the motherboard. The motherboard is made and designed to fit a specific type of chip. For example, most ASUS motherboards are made to fit Intel chips.

The CPU, as many of you may have learned in school, is the brain of the computer. It is responsible for interpreting the code. It makes the code usable for the other components of the computer. The speed of the processor determines the overall speed of the computer. The CPU contains the control unit and the ALU (arithmetic logic unit).

Let us learn about the hard drive. The hard drive, also known as a hard disc, is an essential part of a computer that stores the code that has been processed by the CPU. The hard drive capacity is measured in GB (gigabyte) on most hard drives, but on big hard drives, it is measured in TB (terabyte). 1 TB = 1000 GB

Now, let us learn about the monitor. The display is very important as it helps us see what is going on in the PC. How many frames a monitor is able to play in a second is called refresh rate which is measured in Hertz (Hz).

Another type of memory is RAM, also known as instant memory. If you run many programs at a time, you may need to upgrade your RAM. A computer with more RAM has more multitasking capability.

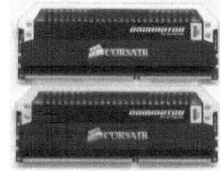

25

Lifespan of a PC

The lifespan of a PC depends on many factors. First, the quality of the computer is a very important factor. Second, how much you use and how well you take care of it are other important factors that determine the lifespan of your PC. If you take good care of it and use good-quality parts, your PC may very well last over five years. But if you do not take proper care of it and use knock-off parts, your PC would be lucky to last three years.

Tips to Ensure Your Computer Lasts Long

1. Do not keep your PC on the floor.
2. Do not keep your PC in the corners, near the curtains and furniture or any other obstruction as this will lead to reduced airflow and overheating which will in turn shorten the lifespan of your computer.
3. Clean your PC regularly using a can of compressed air to remove dust and dirt that might be lodged in the air vents of your computer. If your computer is neglected for a long time, dust and dirt will accumulate and block the vents and overheat your computer, reducing the lifespan.

4. Keep your computer at room temperature whenever possible. Avoid areas with high humidity and extreme temperatures as this will cause your computer and its components to wear quicker.
5. Install an antivirus software and update it regularly. Install windows security updates whenever possible as this will increase the speed and security. Delete useless files to make your computer function efficiently.

If you use good hardware and follow these tips, your computer can last well over six years.

26

Interesting Facts About Computers

Even though laptops are becoming more and more popular, there is still a big market for desktop PCs. A PC takes care of your personal needs. With the help of computers, you can browse the World Wide Web, watch videos and access social media. The possibilities are endless.

According to statistics, the radio got 50 million users in 38 years, but it took only 5 years for the internet to reach 50 million users. **(26.1)** The UNIVAC I was the first commercial computer invented by John Mauchly and J. Presper Eckert in 1951. It had only a very tiny fraction of the computing power of a modern cell phone, but it occupied the entire space of a room. This shows how computers have evolved and developed from bulky underpowered machines to lightweight incredibly fast devices.

The first mouse was invented by Douglas Engelbart in 1968. They have come a long way from the old, bulky ones of the olden days. In 1985, Windows 1.0 was released.

In 1990, Tim Berners-Lee first coined the term World Wide Web. Tim Berners-Lee is also known as the father of the internet. He implemented the first successful communication between an HTTP (hypertext transfer protocol) client and a server over the internet. **(26.2)**

In 1984, the most advanced technology in computer science was introduced and that saw the Macintosh line of computers, still manufactured and developed. The Macintosh was so revolutionary because of its new user interface which is called graphical user interface (GUI). This is the type of user interface that we are familiar with now. Computers pre-dating the Macintosh had a text interface which was often complicated and difficult to operate.

A Ti-83 calculator has six times the power of the computer that operated the Apollo 11 which put men on the moon.

27

Computers and Technology – Questions and Answers

Question 1: What is the difference between gaming computers and normal computers?

Answer: Gaming and normal computers are not that different, except the fact that the gaming computers are usually faster and can handle gaming which requires good hardware. Do not mistake gaming computers for consoles. Some examples of consoles are Sony PlayStation, Microsoft Xbox, Nintendo Wii and Nintendo Switch, all of which have controllers that are equivalent to the mouse and keyboard of a computer. Now PlayStation 5 has been announced along with Xbox Project Scarlett. The difference between a gaming computer and console is that a console is usually cheaper and easier to configure. Gaming computers are usually for gaming enthusiasts. Casual gamers usually prefer gaming consoles.

Question 2: Does the size of a computer matter?

Answer: The size of a computer does not really matter, because even when it is more compact, it may pack a punch. A bigger computer may have more airflow but not in all cases. All the companies are trying to get their computers more compact to save space. So, the size of the computer does not really matter.

Question 3: Which is preferable—a pre-built computer or an assembled computer?

Answer: We cannot say that any one of them is a better choice because it is a matter of preference. If you want to save some money and have some spare time and believe that you have enough skills to assemble a computer—which might seem to be hard—assembling a computer by yourself is the way to go. It must be noted that assembling a computer is not as hard as it looks. You can do it by looking at a simple tutorial on YouTube. But if you want a computer with no tweaking to do and you have extra money to spend, pre-built computers are the way to go. I cannot say that this one is better than the other one. It comes down to preference.

Question 4: What is a pop filter? What is the purpose of it?

Answer: A pop filter is a noise reduction device. The pop filter's purpose is to stop air from going through the microphone so that it does not pick up the noise of air or the rattling noise or the noise of the fan. It allows you to get a smooth recording. Pop filters allow you to have your fan or noisy air conditioner on while recording audio.

Question 5: Which type of camera is good for YouTube gaming?

Answer: I would recommend a webcam that can be mounted right on top of your monitor rather than a big camera that cannot be mounted on the monitor. If you prioritize quality, you can go for a big camera that can be connected to the PC so that the video quality is very good.

Question 6: Do you really need to configure your gaming peripherals before using them?

Answer: No, not really. But, if you need your peripherals to be fancy, you can configure them to make their performance slightly better or to tweak some settings. For example, tweaking the DPI (dots per inch) on a DPI customizable mouse.

Question 7: What is the best free software for making YouTube videos?

Answer: The best free software for recording and producing YouTube videos is Open Broadcaster Software (OBS). It is for the most part easy to use, and if you want to be professional, you can tweak the settings for maximum results.

Question 8: Does the screen resolution really matter?

Answer: The screen resolution affects the performance of a computer a little bit. A larger resolution requires more processing power; it may reduce FPS (frames per second) but it produces high-quality, crisp image. Lower resolutions are not very nice-looking, but they increase FPS and performance.

Question 9: Which gaming company makes the best mice?

Answer: Two main gaming companies make very good products and are popular as well. Like some of the answers to other questions, this also boils down to preference. Those two companies are Corsair and Razer. Razer makes very beautiful and appealing high-quality mice. Corsair does make very appealing mice, but they focus mainly on performance. If you want a fancy, high build quality mouse, choose Razer. If you want a high-performance mouse, go for Corsair. It must be noted that both Corsair and Razer mice are costly. But the low-end of Corsair and Razer mice range from Rs 3000 to Rs 4000 while their high-end mice go from Rs 33,000 to Rs 40,000 for Razer and Rs 20,000–Rs 22,000 for Corsair.

28
Artificial Intelligence (AI)

Introduction

AI is a component of computer science. It is also known as machine intelligence. AI is intelligence manifested by machines that perform some of the activities of humans.

Let me provide an overview of how AI works. AI learns, collects gigabytes, terabytes of data. Then AI analyzes the data using the algorithm and it finds out how to do a specific task. This AI is called weak AI. It needs to be instructed to do any task. It knows it but it needs to know when to use. But the strong AI, on the other hand, does not need to be instructed as it knows whatever it wants to know and it uses it whenever it faces a problem.

We, humans, are like strong AI. We know a lot of information and whenever we face problems in our day-to-day life, we use our intelligence to solve these problems, in the sense that we are like strong AI.

Deep Learning

There have been many types of AI. But the most, common, popular and powerful one is deep learning. What deep learning does is it uses data  and algorithms which humans have fed to it. For instance, if deep learning AI is fed with many pictures and videos of cars being driven, it will learn to drive a car from the tens of thousands of pictures and videos fed to it.

An example of deep learning is face-swapping. In face-swapping, you can take either your face or someone else's face and swap it with another person's face in a video. This is done using deep learning AI. It tracks your face's motion and makes a three-dimensional model of your face and then it swaps it with the face of another person. Deep learning is a deeper, more complicated neural network with more layers.

Artificial Neural Networks

Artificial neural networks are a type of AI that is used for many tasks. One of them is automating playing video games. The example of video games is easy to understand, so I will use that as an example. Artificial neural networks achieve this by trying repeatedly. Every try or attempt at reaching the goal is known as a generation. The successful attempt or the attempt that is closest to success will move on to the next generation and after 20–1000 attempts, depending on the task, one member of the generation or many members will succeed in the task.

AI is as complicated as the human brain. By the way, you can find a lot of information in "How Does Our Brain Store Memories?" Here, how memory is stored is not so complicated as how we process the data.

I have been intrigued by robots since my childhood. And it never struck my little mind that this would be transformed into such a great phenomenon. Now people are scared that it will replace human jobs. But the matter of the fact is that creative jobs, such as a CEO, an artist, a singer and a dancer or anything of the creative artists cannot be taken by robots. Robots may have logic, reasoning, intelligence, but they cannot replace creativity or generate ideas. Robots do not have free will, choice-making ability and decision-making skills. Moreover, they do not have feelings. In order to socialize, robots need human beings. Robots have taken away and are also going to take away routine and repetitive jobs of people like truck drivers, drivers, shoppers and chefs but they cannot take away jobs that require creativity, love and care, and patience. Robots will never and can never replace love and creativity because they do not have life or emotions. We have life. Artificial life is not the same as real life. They will replace probably all the routine and repetitive jobs, leading to a decrease or total destruction of these jobs, but the demand for creative jobs will increase.

It is said that even robots can become surgeons, but the fact is that they can never give the care and warmth that humans do give, yet AI is so fascinating.

AI is learning and keeps on learning. AI learns from its surroundings, and at one point, it will exceed our intelligence. The evolution of AI is scary but fascinating. It has opened doors for great opportunities, but has also closed the doors for jobs. It reduces jobs in one sector but increases jobs in another sector. It is like you lose one and take two, but people are always scared about losing that one thing. These are the basics of AI.

Applications of AI

AI is a powerful tool. The development of AI is one of the greatest achievements of humanity and its power is not to be underestimated. Let us look at the applications of

AI. Nowadays, there are so many algorithms and natural language processing AI. Many business establishments are using these technologies to bring down their machine load and make their workflow more efficient. Companies that use AI profit much more than those which do not use AI, by cutting down on their machine load and maximizing the efficiency.

1. Fraud Detection

The use of AI in fraud detection in banks is highly beneficial. Banks view a lot of normal transactions, for example, a normal exchange of cash. It knows what is normal and whenever it detects an abnormal transaction, it alerts the person concerned who looks into it and finds out whether it is actually abnormal or not. Hence, he does not need to go through all the normal transactions. This is very useful for banks that cut down some significant workload.

2. Online Customer Support

Whenever you go to Amazon or Flipkart or Myntra or any other online services, you may have noticed a customer helpline rather than a human taking the call to your aid. The AI first takes the call and it helps you with the data gathered from the website of the company. When you go to Amazon, for example, if you do not know the procedures of online banking, you call up the AI you have. When you ask the AI "How should I choose my payment mode?" it responds to your inquiry by assisting you with the data it has gathered from the website of the company.

3. Cyber Security

Cyber security is improving much more than the advanced technologies that the hackers use. Big firms are investing in cyber security to ensure that their data is protected and secure because data is invaluable. Information is money. Threat removal and breach prevention are what companies are looking for and the solution to all of these is the application of AI.

Using machine learning algorithms, security experts can teach AI how to analyze and ultimately prevent breaches and hacks and all of these are just the tip of the iceberg.

Humans' Harmony with AI

Many people are worried that AI will take over the human race and the reason attested to the fact is that it has already taken away many jobs. Yes, it is really a big issue. It is assumed that AI cannot supersede or supplant humans.

AI cannot replace human companionship and the feeling of human love and the attribute of human behavior. Robots or AI can never replace executive jobs and jobs that require productivity. AI cannot also take the leadership of humans; they cannot be CEOs of companies. It is impossible. Humans are made for humans. Humans can best attract humans and not robots. Let me illustrate this with a story. Once there was a man who wanted to create a very simple device that everybody could use, for example, calling their doctor, ordering food or calling the customer helpline. It was a success. They checked the statistics and found out something shocking. Most of the people who contacted the customer helpline told them that they called because they felt lonely, sad and depressed. They did not want to either call their doctor or order food, but all that they wanted was someone to talk to. **(28.1)** I can talk to you. Do you want AI to talk to you? That companionship AI misses.

Humans have been endowed with the faculty of writing books. I have written a book that you are reading now. AI also can write books. But AI will be an inferior writer because often it lacks an interesting plot and its writing will be repetitive. AI cannot also create new book ideas. It can only continue a series of books already started. Therefore, a book written by a robot will be uninteresting and boring.

There are great things that robots have done, for example, they have done surgery on a grape. It is so amazing to see a surgeon operating the robotic arm with such precision. Robots can replace a driver, a surgeon, and so let us go with them, use them and be in harmony with them rather than competing with them. Instead of considering them as our rivals, we should make them our partners. But there are some dangers of AI which I have discussed in the next section.

AI's Threat to Human Race

With AI's potential growth at an unprecedented rate, we cannot help but wonder about the negative effects of AI. To understand the potential dangers of AI we have to establish three types of AI.

1. ANI (Artificial narrow intelligence)
2. AGI (Artificial general intelligence)
3. ASI (Artificial super intelligence)

Currently, we have achieved only ANI. ANI is good (not better than a human) at a specific task but not good in other tasks. We have not achieved AGI or ASI. AGI will be as good as a human or sometimes better than a human at all tasks. ASI will be better than humans at all tasks. It will only take less than a day for AGI to turn into ASI, as it will teach itself about how to get better at teaching itself.

One of the things that AI can do that humans cannot is fundamentally rewiring its brain in order to learn things better. AI considers humans as a hindrance to its development. Great men such as Stephen Hawking, Elon Musk and Bill Gates think that AI poses a threat to the human race itself. Stephen Hawking, a renowned theoretical physicist, thinks that AI will either be the worst or best thing to humanity. Elon Musk who owns an AI company thinks that the power of AI should not be concentrated. The reason is if a person with bad intent steals the AI, he will obtain unimaginable power. To avoid such a catastrophe, Elon Musk founded OpenAI, a company dedicated to making top-level AI open and accessible to the public. Bill Gates also thinks that AI poses a great threat to humanity. Many more people talk about the dangers of AI. Experts estimate that it will take us a long time to make AGI which will quickly turn into ASI. The great minds Stephen Hawking, Elon Musk and Bill Gates presume that AI is dangerous. **(28.2)** As we always have to have a leash on our dog, we should always have a leash on our creations. I think it is worth keeping an eye out for AI.

Conclusion

AI works by collecting huge amounts of data, and it is really helpful. We have seen many uses of AI in cyber security, fraud detection and customer support. Robots can never replace humans as humans are superior to robots. Though AI poses a great threat to humanity, instead of considering it our rival, we should have harmony with AI. When we have harmony with AI, it can open new vistas and do great marvels that can shoot the human civilization to an unimaginable extent of technological innovations. I look forward to seeing an amazing future. We can create robots that can help people worldwide. They can save lives and make life easier in spite of their limitations. Only when we use AI properly, that is, when scientists and technologists are using AI constructively, the world in which we live will be a heaven on earth.

29
What Have People Gained by Watching and Making Tech YouTube Videos?

People have achieved many things by watching and making YouTube videos.

First, let me list a few benefits of watching YouTube videos. Through YouTube, you can get the latest news. You can gain access to one of the biggest media libraries on the internet. YouTube also gives you a platform to express your views. You can learn complex concepts in an easy way on YouTube. I have learned English by watching English YouTube videos. You can learn so much if you watch educational YouTube videos. YouTube is the place to go if you want updates on the latest and the greatest technology. YouTube also has a lot of product reviews. If you are looking forward to buying something, you can look up reviews for it and get all the information you need before buying the product.

Here is a brief account of the benefits of making YouTube videos. Making YouTube videos has been a lucrative profession for many people. YouTube pays you if you enter the YouTube Partnership Program. People are paid based on the CPM (clicks per thousand, M is used for thousand as it is the Roman numeral for thousand). Top YouTubers are paid a lot of money for their videos through advertisements and sponsorships.

Marques Brownlee is a tech reviewer and YouTuber. He started making videos around the late 2000s when he was very young. He did not have many subscribers for a long time until one of his videos blew up and catapulted him into fame. Since then he has been consistently making videos and he has sustained his growth. Currently, he has around 8.9 million subscribers and the number is growing. He became so famous through his videos that led him to interview a tech tycoon and philanthropist Bill Gates, a tech legend Elon Musk, a renowned basketball player Kobe Bryant and an admirable astrophysicist Neil deGrasse Tyson. He met and talked with all these people only because of the fame that he had garnered through YouTube.

The next famous YouTuber is Linus Sebastian who is the man behind Linus Tech Tips, TechLinked and Techquickie. He started out working in a computer shop and making videos for it. He later quit the job and started his own channel to make his own YouTube videos. He started the Linus Media Group. His channel grew rapidly and now he has his own studio and office. He has collaborated with many other big YouTubers to make videos. He has even collaborated with Marques Brownlee.

30
How to Make YouTube Videos

Many YouTubers have different methods of making YouTube videos. I am going to mention only the most popular methods that are used by renowned YouTubers. It all starts with the idea. Many YouTubers have different ways of generating ideas. The idea is very important. It is also important to form the fundamental structure of the video. The structure is very important, as the whole video needs to be built on it. The structure will provide a sort of narrative to the video. A good idea is important for getting people interested in the video.

A trending topic may also garner views. But, milking the trends will often make loyal fans dislike the YouTuber. Riding the trends will also build an ever-changing and fickle audience but it may not be good in the long run. Building a loyal audience, that is, an audience who will be ever eager to watch your videos is the key to success on YouTube. Technology is a hot topic because there are many things to talk about. You can build a loyal audience and get views from other YouTubers.

The idea is only the beginning. Execution is very important. Once a person clicks on the video, you have to make that person keep there throughout. The quality of the video is very important. You need some basic gear. With some basic, cheap gear, you can make great videos if you have the passion. But if you have money to spend, good gear can improve the quality of your videos. Here are the basics: you need a microphone, pop filter and a webcam or any camera. If the computer that you use has a microphone, you can use that. For good quality, you have to use an external microphone. Popular YouTubers use a lot of technology to make good-quality videos. They have huge recording studios and perfect lighting. They usually use external video cameras to get good video quality. RED cameras are the cameras of choice for YouTubers. They are expensive and have amazing video quality. If you do not have a budget for an external camera, you can use a Logitech webcam. Many YouTubers have expensive microphones for great sound quality. Audio-Technica is a popular microphone company. Its microphones are of good quality. They also have studio lighting. Having studio lighting is not very important. Make sure you have good room light or sunlight.

Once you have recorded the video, you have to edit it. Editing is an art. Editing is an important part of video production. You can use editing to tell a story. There are many editing software for you to use; many of them are free. People who have experience in editing can use DaVinci Resolve. People who are not so experienced can use OpenShot. If you have some money, you can use Adobe Premiere Pro. I would highly recommend Adobe Premiere Pro. If you want something even simpler, you can check out Sony Vegas Pro. Buy only Vegas and Premiere Pro if you want to do some complex editing.

I list some websites for you to get help on these editing software:

OpenShot:

www.openshot.org/user-guide/

Premiere:

helpx.adobe.com/in/premiere-pro/user-guide.html

Vegas:

https://www.vegascreativesoftware.com/in/downloads/

DaVinci Resolve:

https://documents.blackmagicdesign.com/UserManuals/DaVinci_Resolve_14_Reference_Manual.pdf

After editing this, you have to make a thumbnail and publish the video. A catchy title and a good thumbnail are important to get a good number of views. You can come up with a creative video title. The thumbnail can be made through websites like canva.com or programs like Adobe Photoshop. If you follow all these steps and tips, you can make a great YouTube video.

Now I am going to tell you how to make a YouTube account. To make a YouTube account, you have to have a Google account. You can make a Google account by signing up for Gmail. To create a YouTube channel, go to https://www.youtube.com/create_channel and enter all your credentials. To upload a video, you have to know where the upload button is. On the top right corner, hover

over the buttons and click the one which says "create a video." The icon could change anytime.

The YouTube realm is an ever-changing one. So, it is important to understand all the video types on YouTube. There are many tech videos. Comedy is also a popular genre on YouTube. Some people vlog. Vlogging is video blogging. A blog is defined as a regularly updated website or web page, typically one run by an individual or a small group, which is written in an informal or conversational style. Some people do commentaries. Some people game. There are so many video types to choose from. You do not need expensive software and hardware to make amazing videos. If you have passion and creativity, you can create a stunning video. You can use YouTube videos to tell a story, to educate and to entertain. If, at first, you do not see any progress, you should not lose heart. If you keep making videos with the same passion and creativity, you can earn a lot of subscribers. The person who is the most subscribed YouTuber of all time had, at first, only 200 subscribers in a year. Now, after ten years, he has 90,000,000 subscribers. Keep experimenting with video types to see what works best. That is the best advice I can give you.

31

How Do Websites Store Your Password?

Have you ever wondered where your password goes? How do websites store your password? What is the safest way to store passwords? How can we protect ourselves against hackers?

Method 1: Plaintext

The first and the most naive approach is simply storing the passwords in the database as plaintext without doing anything to them. The database stores them as plaintext indicating this is the username and this is the password. But it is unsafe, because if a hacker or an insider gets access to the database, he can get access to all the passwords which are stored inside it. So, this is a very naive and very insecure method of storing passwords.

Method 2: Encryption

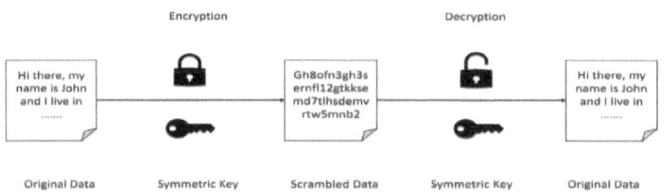

Now, let us discuss briefly how encryption improves security. Encryption is a more secure method of storing passwords. There are different types of encryption. First, we will discuss the simple one. The computer would store the username as it is, take the password and put it through an algorithm that would change the password into a different language, not from Tamil to English or Hindi to English but to the English alphabet only, but in a weird convoluted way. It puts it into an algorithm, and the algorithm changes it into something else. This is a secure approach. The database may store it in a hash.

Second, there is a strong type of encryption called 256-bit encryption. Google uses it and it is more secure.

Method 3: Hashing

What hashing is that it takes the password, mixes it up and locks it behind a vault. And when you enter your password, it acts as a key and unlocks the vault. It compares the key (the password you have entered) with the hash and it allows access, only if the hash and key match.

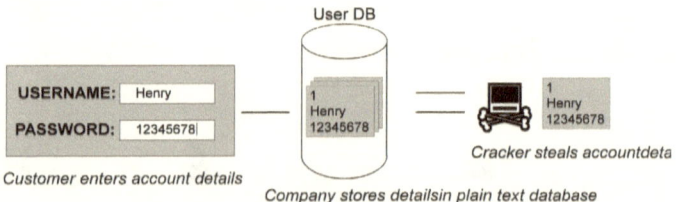

Customer enters account details

Company stores detailsin plain text database

Cracker steals accountdeta

Hashing is a very secure approach if done right.

But the problem with hashing is that when the door is open, or when the vault is open, hackers can get into it. This makes it a little bit insecure.

Method 4: Hashing and Salting

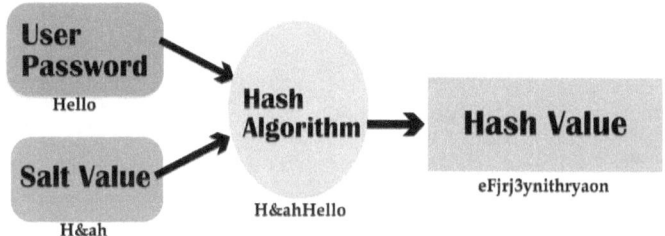

Now, let us see another method that is more secure. You hash it, encrypt it and sprinkle a little bit of salt on it. This method is called "salting" in the metaphorical sense. The algorithm generates random letters like a c d j b j m and it puts the string of letters and mixes them with the final code. So, you get the diluted form of the original password which is very hard to guess by computers. Only one thing you can do, that is, just guessing the password. This is the best approach. Adobe, a company that made Acrobat and Photoshop, once did not store its passwords properly. This resulted in a big data breach which led to a lot of passwords being leaked.

Here are a few tips to keep your passwords secure:

1. Do not use the same password for all services.
2. Do not use very common passwords.
3. Change your password often.

4. Use a passphrase with numbers, capital letters and small letters.
5. Do not share your password.

(Source: 31.1)

32
Copyrighted Numbers

Let us explore why some numbers are copyrighted. A copyrighted number is a number you cannot distribute. Copyrighted numbers have a long history.

To understand copyrighted numbers, first, we have to learn about prime numbers. Prime numbers are numbers that can only be divided by themselves and one, and they do not get a quotient. A prime number is also defined as a number that has only two factors (the number itself and one). Computers can easily multiply two big prime numbers. Let me use two small prime numbers as an example: 17, 27. A computer can calculate this as 17×27 = 459. But it is very hard for a computer to calculate the factors of 459. Computers, given some time, can work the factors of 459. Now, they use 30–100 digit primes. It will take even for a supercomputer a long time to work. That is why people use prime numbers to encrypt DVD's anti-piracy measures. Moreover, the distribution of the prime numbers used to encrypt DVDs is prohibited and it can get anyone arrested in the USA. If this prime number is released to the public, people can get paid music for free because they have cracked the encryption of the CD using this number. This makes it the ultimate copyrighted number. This raises a few questions. Can you make a copyrighted number? What are the other things that can be copyrighted?

As CDs declined in prominence, the copyrighted numbers faded away. Currently, only the prime numbers that are used to crack normal encryptions are copyrighted. Will the cybersecurity experts crack down on these numbers? Only time will tell.

(Source: 32.1)

33

The Most Influential Innovators and Their History

Bill Gates

Bill Gates is one of the most influential people on the planet Earth. He is the second richest man in the world, next to Jeff Bezos. Bill Gates is the founder of Microsoft. He was born in 1955 in Seattle, Washington, USA. He founded Microsoft along with his childhood friend Paul Allen in 1975. Bill Gates and Paul Allen made Traf-O-Data, a machine that would count traffic, before founding Microsoft a couple of years later. According to Bill and Allen, the Traf-O-Data project helped them gain experience in writing the Altair BASIC code. Unfortunately, Paul Allen passed away in October 2018 due to cancer. Bill Gates relinquished the position of CEO of Microsoft in 2000. Gates is now an active philanthropist and humanitarian. His philanthropic efforts include malaria prevention and poverty reduction. Their beliefs are found in the Bill and Melinda Gates foundation's website https://www.gatesfoundation.org/. **(33.1)**

Here is an excerpt from the website: "We work with partners to provide effective vaccines, drugs, and diagnostics and to develop innovative approaches to deliver health services to those who need it most. And we invest heavily in developing new vaccines to prevent infectious diseases that impose the greatest burden."**(33.2)** Their efforts have saved thousands of lives. They also do a lot of work in bringing about awareness of deadly diseases prevalent in developing countries. They also provide education and employment for women in developing countries. They also work with local communities in order to boost the local economy in those countries. Their work has made the world a better place. Aside from philanthropy, Bill Gates is one of the greatest innovators in technology.

Steve Jobs

Steve Jobs was born in San Francisco in 1955. He was adopted by Paul and Clara Jobs. He co-founded Apple along with Steve Wozniak and Ronald Wayne. Apple is famous for a number of products, including the iPhone, iMac, iPad and MacBook. Steve Jobs, unfortunately, passed away in 2011 due to pancreatic cancer. **(33.3)**

Steve Jobs and Steve Wozniak built the world's first widespread personal computer called "Macintosh." Even today the Macintosh is manufactured and updated. During Steve Jobs' tenure as the CEO of Apple, he brought about many innovations that changed the world. For example, he introduced the iPhone, that is, the smartphone that sparked the "smartphone revolution." The iPhone catapulted Apple to new heights. One of the other innovations that Steve Jobs introduced was the iMac which was the continuation of the Macintosh lineup of computers. The iMac was acclaimed for its sleek design. One of the unique features of the iMac was that all the parts of the computer were located in the monitor itself like the original Macintosh. The iMac was very popular with designers who liked its sleek design. One of the other innovations brought about by Steve Jobs was the introduction of the iPod. The iPod was an MP3 player like no other in its time. It was compact, portable and convenient. It was popular with runners as it was easy to carry around during exercise. He resigned shortly before he passed away. Currently, Tim Cook is the CEO of Apple.

Jeff Bezos

Jeff Bezos was born in 1964 to Miguel and Jacklyn Bezos. He is well known for founding the very popular online store Amazon.com. He is also the richest man in the world followed by Bill Gates. His enormous wealth is due to his holdings in Amazon.

Amazon is one of the biggest companies in the world. In 1984, Jeff Bezos created Amazon in Seattle, Washington. Amazon began selling books, music and videos. In 1998, it began operating internationally by acquiring sellers of books in the UK and Germany. From 2000, Amazon began selling a wide range of products. Amazon became popular through its AWS portfolio. Jeff Bezos, through his perseverance, has made Amazon a $700 billion company. Jeff Bezos also owns a space exploration company called Blue Origin. **(33.4)**

Mark Zuckerberg

Mark Zuckerberg was born in 1984 to Edward Zuckerberg and Karen Kempner. Mark co-founded Facebook along with his dorm roommate Eduardo Saverin, Andrew McCollum, Dustin Moskovitz and Chris Hughes. On February 4, 2004, Zuckerberg launched Facebook from his Harvard dorm room. Facebook exploded in popularity. It reached around one billion users in 2013. **(33.5)** He said to Wired magazine in 2010: "The thing I really care about is the mission, making the world open." **(33.6)** He has launched many new ventures, one of them being Internet.org which aims to provide internet access to billions of people who do not have internet facility. He believes in an open world where everyone is able to communicate freely and openly. Mark Zuckerberg is one of the world's top tech entrepreneurs.

Elon Musk

Elon Musk is an entrepreneur and CEO of SpaceX. He was born to Errol and Maye Musk in 1971 in South Africa. Elon Musk is famous for being associated with Tesla, SpaceX and The Boring Company. He founded SpaceX in 2002 and co-founded Tesla in 2003. Elon Musk is the 40th richest person in the world. **(33.7)**

Tesla is the best electric car manufacturer in the world. Tesla is famous for its great innovation in the field of electric cars. SpaceX is a company similar to NASA. The only difference is that SpaceX is privately owned rather than government-funded like NASA. SpaceX is one of the best space transportation service companies in the world. SpaceX is also known for innovation in its field. SpaceX is famous for its landing of reusable Falcon Heavy rocket boosters. Elon Musk's personal car was used as a test payload for the Falcon Heavy rocket which is currently floating in space. His innovation in the field of space exploration is amazing. Musk also owns a solar farm, and he is a big believer in renewable energy. He has also done great work in the field of AI. Musk's AI company is one of the leading AI companies in the world. His space exploration company is one of the biggest in the world and is acclaimed for its contribution to space research.

34

The History of Microsoft

Microsoft was founded on April 4, 1975, by Bill Gates and Paul Allen. Microsoft started out making and selling BASIC interpreters for the Altair 8800 computer. It rose to prominence with the MS-DOS  operating system which was the predecessor to the very popular Windows line of operating system. From the mid-1900s to the late 2000s, Microsoft started to realize

 the potential of the World Wide Web. In 2000, Bill Gates handed over the position of CEO to his college friend Steve Balmer who had been an employee of the company since 1980. Bill Gates took the position of Chief Software Architect. In 2001, the popular Xbox series was launched. The first model of the Xbox was the Xbox 360 followed by the Xbox One in 2013. Bill Gates quit his position as Chief Software Architect in 2008. Windows tried to dip its toes into the smartphone

business in the late 2000s with the launch of Windows Phone.

The Windows Phone failed as they entered the smartphone business late, compared to Apple which launched the iPhone in 2007. One of the mistakes that Microsoft did was it tried to create a closed ecosystem with a phone that was not manufactured by  Microsoft. Though revenue tripled and sales doubled during Steve Ballmer's leadership, he was one of the most hated CEOs at that time. People accused him of only trying to sell products and not innovating anything in a big way.

In 2014, Satya Nadella took over Steve Ballmer as CEO of Microsoft. Satya Nadella, rather than selling products aggressively, tried to innovate. Satya Nadella was the leader of the Cloud Computing Department prior to becoming the CEO. In April 2014, Microsoft acquired Nokia for $7.2 billion. Nokia was made a subsidiary of Microsoft and renamed as Microsoft Mobile. Microsoft's mobile division is now dead and does not manufacture phones anymore. Microsoft has also discontinued software updates. Microsoft has made some innovations recently.

In 2016, Microsoft unveiled the HoloLens 1, a holographic lens that works similar to a Virtual Reality (VR) headset. It allows the user to interact with the holograms that are projected. The HoloLens came with a hefty price tag of $3000 or Rs. 2,09,265. It was targeted at designers and  other professionals. There is a new iteration of the HoloLens named HoloLens 2. The HoloLens 2 was released in 2019. HoloLens 2 is much more refined and has better display capabilities and a wider field of view.

Microsoft produces a huge lineup of products ranging from gaming consoles to holographic lenses. Microsoft is now one of the biggest companies in the world.

(Source: 34.1)

35
The Most Innovative Software and Websites

1. Facebook

Mark Zuckerberg founded Facebook while studying psychology at Harvard University. Membership was initially restricted to students of Harvard University. Later, Facebook spread its tentacles to other parts of the world. Facebook is a social  networking site founded in 2004. Currently, Facebook has close to three billion users. Facebook went public in 2006. Anybody above 13 years can sign up for a Facebook account. Many people assume that Divya Narendra is the founder of Facebook. This is not true. Divya Narendra is the founder of Harvard Connection, not Facebook. Facebook grew rapidly in 2010. It has millions of people around the world as its users. Facebook claimed the throne from another popular social networking site, Myspace. Facebook's success is unparalleled, even when compared to other social networking titans, such as Twitter and Instagram. Facebook's success made its founder Mark Zuckerberg the eighth richest man in the world as of 2019.

2. Adobe Photoshop

Adobe Photoshop is a raster editor developed and published by Adobe Inc. for Windows and macOS.

A raster graphics editor is a computer program that allows users to create and edit images interactively on the computer screen and save them in one of many "bitmap" or "raster" formats, such as JPEG, PNG and GIF. It was created in 1988 by Thomas Knoll and John Knoll. The Knoll brothers sold Photoshop to Adobe Inc. in 1988. Photoshop 1.0 was released in 1990 exclusively for the Macintosh. Photoshop is one of the most successful programs of Adobe to date. It was and it is the best image editor. It has a lot of fine-tuned and polished features and user-friendly interactivity which makes it widely popular. Many people have made art with Photoshop. **(35.1)** Editing photos gets easier day by day.

3. Google.com

Google.com was launched in 1998. It was founded by Larry Page and Sergey Brin.
It is the most used search engine on the World Wide Web across all platforms, with 92.74% market share as

of October 2018. Almost everyone uses Google. Google has evolved from a search engine to a company. Google.com has a monopoly on the search engine field. Google.com has innovated the way people traverse the World Wide Web. Google has provided us with an easy and quite convenient way to search the World Wide Web. Without Google, much of the World Wide Web would be unknown to us.

4. YouTube

YouTube.com is a popular video hosting site. It was founded in 2005 by former PayPal employees Chad Hurley, Steve Chen and Jawed Karim. Google bought YouTube in 2006 for $1.65 billion. YouTube has been operating as a subsidiary of Google since 2006. Most of the content on YouTube is uploaded by individual creators or indie creators. But, media corporations including CBS, the BBC, Vevo and Hulu offer some of their materials via YouTube, as part of the YouTube partnership program. YouTube creators earn money on the AdSense program. YouTube creators earn money based on CPM (cost per thousand, the M stands for thousand). It is the cost to display an ad (advertisement) to a thousand people. YouTube has given millions of people a platform to make content and express themselves. Most videos on YouTube are free to watch. But YouTube also offers a premium subscription.

5. WhatsApp

WhatsApp, also known as WhatsApp messenger, was released in 2009. It is a Voice over IP or VoIP service. Facebook owns WhatsApp. WhatsApp was bought by Facebook for $19.3 billion. By February 2018, WhatsApp had 500 million users. It is very popular in India, Brazil and most parts of Europe. WhatsApp has exploded in popularity in India. WhatsApp has 200 million users in India. WhatsApp supports ten Indian languages. Fifty-six percent (56%) of people in India who have internet connection use WhatsApp. The country that tops the use of WhatsApp video calls is India. WhatsApp is equally as popular as Facebook in India. WhatsApp is used in 180 countries. WhatsApp has connected people all over the world.

36
Cloud Computing

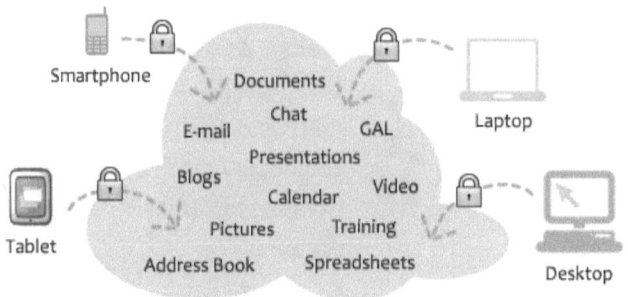

Cloud Computing
Having secure access to all your applications and data from any network device

Cloud computing is commonly defined as the practice of using a network of remote servers hosted on the internet to store, manage and process data, rather than on a local server or a personal computer. Cloud computing allows its users to have great computing power without building a computer of that power. The computer's "power" is accessed over the internet.

Many of you may be familiar with cloud storage and cloud storage services like Dropbox and Google Cloud. Services like these store the information or data in servers and data centers around the globe that can be accessed over the internet. Cloud storage allows the user to store huge amounts of data safely and without the complication of a local server. Cloud storage is relatively inexpensive compared to building your own server which requires expertise and a lot of money. Google Cloud and Dropbox are free up to 15 GB. After 15 GB, if you want more storage, you can pay for it. It costs around Rs. 70 for 10 GB for one month with Google cloud, which is very cheap for that kind of storage. Google Cloud also offers cloud computing solutions. Another popular cloud  computing company is Amazon Web Services (AWS) which is a subsidiary of Amazon. Both Google and Amazon offer similar price.

An exciting innovation in cloud computing is cloud gaming. Cloud gaming has been around since 2012. But, only recently in 2019, the concept of cloud gaming started gaining traction with Google announcing its take on cloud gaming called "Google Stadia." As I have mentioned before, cloud gaming is not a new concept, but Google has a lot of cloud computing infrastructure and funding to bring cloud gaming to the mainstream gaming market. Cloud gaming allows the user to play any video game online over the internet without downloading it. Google might face some issues. Latency is a big issue with respect to cloud gaming. Latency is basically a delay due to poor internet connection.

Most cloud gaming services are not free. As of 2019, 99% of cloud gaming services are paid. They are not very expensive. There are only a few free cloud gaming services, one of them being GeForce Now. GeForce Now is in free beta, which means it is in a stage of development and it is free.

One issue with GeForce Now is that it is an invitation-only which means you have to wait in an internet queue to download it. The queue may be months long. Unfortunately, GeForce Now is not available in India. Cloud computing has great potential. Cloud computing holds the key to the future. I am excited to see future cloud computing innovations.

37

Computers and Gaming in Today's Modern World

In the beginning, when computers were introduced, they were usually big and boxy. Nowadays, they are compact and sleek. Usually, compact computers are more expensive than big computers. Before learning about gaming computers, let us know more about the CPU.

The CPU is a chip inside the computer. Some people think that the whole computer is the CPU, but this is a misconception. Already we have seen that the CPU is the brain of the computer. The other things, such as power supplies, graphics card and cables are not parts of the CPU. A device called "heatsink" is mounted on top of the CPU to cool it down and prevent a phenomenon called "thermal throttling."

Now, let us learn about gaming computers. The prefix "gaming" is used to describe high-end computing equipment that can handle gaming. Mouse, keyboard and headphones are known as peripherals. Gaming computers need high power so that they do not lag.

Due to inadequate hardware, the game's code fails and becomes unresponsive or crashes (closes) and this is called "lag." This is caused due to either slow internet speed or no internet connection for online games or inadequate hardware.

When the problem is not due to hardware, but the game itself, it is called a "bug." Sometimes, when bugs are exploited, the player can gain an unfair advantage called "exploits" and the action of using exploits is called "exploiting."

FPS denotes how many frames are being played in a second. FPS is dependent on hardware. If you have a good computer, you will have higher FPS as your computer has more processing power. If you have a low-performance computer, you will experience low FPS and lag. Low FPS makes your game slow and your gaming experience bad. One way to get more FPS is to sacrifice the quality of the game for more FPS. If you want high-quality gaming with high FPS, gaming computers are the way to go.

The mouse is very important to the gaming experience. It is one of the main input devices in gaming. There are two main types of modern mice: optical mice and laser mice. As technology has advanced, both laser and optical mice are evenly matched on performance. The only difference is that the laser mice use visible light, whereas the optical mice use infrared light.

Let us learn about keyboards. Keyboards are very important to both gaming and typing. Gaming keyboards feature specialized key switches for maximum response time which is extremely important in gaming. They usually have fancy lights for a cool effect.

Now let us learn more about monitors. Monitors are very important, as they allow you to see where you are and what you are doing in video games. To ensure high responsiveness, monitor manufacturers are working on a real-time monitor which will show the information as soon as it is received. There is also another thing known as refresh rate which is measured in Hertz. The Hertz rate indicates how many FPS your monitor can handle.

38
The History of Video Games

Video games have been around for 60 years. How have they changed during these years? In 1958, the first video game was invented by William Higinbotham who was also part of the team that invented the nuclear bomb. The video game took only a few hours to design and it was a huge success.

In the 1960s, the video game industry was dormant. But, the video games saw a resurgence in popularity in the late 1970s and the early 1980s. Early arcade machines popped up in the 1980s and home consoles were also becoming popular. Video games like Galaxian, Defender, Pac-Man and Donkey Kong became very popular. Pac Man became a cultural phenomenon in the USA. Home consoles became extremely popular in the early 1980s. Companies such as Atari and Nintendo were at the forefront of this revolution.

The rapid growth of the industry was not sustainable and as a result, the industry crashed during "The Video Game Crash of 1983." At the same time, home computers started to take off and hobbyists wanted to play video games on the computer. Game developers found the code and published it in books and magazines. People typed in the code in order to play the game. That is not possible nowadays as the code for modern games is hundreds or thousands of pages long. In the 1990s, 3D video games started taking the front stage. Consoles such as the PlayStation made 32-bit gaming widespread. Graphics have improved a lot. Games are now huge in size and have 3D graphics. Many notable games of the past are fun to play even now. There are still popular game series that have new releases today.

There are many genres of video games, such as Sports, Racing, Open World, Sandbox and Indie. Let me give you some examples of these genres. There are many sports games, such as EA's NBA 2K, Madden NFL and the "skate" series. All these games were developed by EA (Electronic Arts). EA is one of the most popular sport game developers. There are different racing games, such as Forza Horizon, F1 2018 and Copa Petrobras de Marcas. Some examples of open-world games are The Legend of Zelda and Spider-Man. A popular example of a sandbox game is Minecraft. One of the most popular Indie games is Undertale.

(Source: 38.1)

39

Video Games – Good or Not?

Video games–Are they good or not? The answer is both. They can be good like Minecraft for which you eat your heart out. That is what people like. There are violent games as well. So, I suggest you see the ratings of everything. This is just a warning to all the parents and kids out there. I think you should all be aware of what you are buying and what you are playing.

Now, let us get to the critical point of the issue: Are video games good or not? I cannot say "Yes" or "No." Some video games can be good and some bad. Violent games and shooting games are not good for children. I think some video games do help improve your memory. Here, I would like to recount my experience. I started playing video games when I was three years old and they were pertaining to mathematics. It is my mother who taught me to play video games. You can play video games. But, you should be very cautious about what you are seeing, what you are doing and what you are going to do. Enthusiastic gamers like you and me get on with our games and play. At the same time, we should see that they have good content. We do not want to see bad content. Some games that I suggest to the children of all ages are board games, such as Monopoly and The Game of Life. They are physical games. You can play them and you should check them up. There are hundreds of video games that are suitable for all ages. My favorites are Minecraft, Kerbal Space Program and skate3.

40

Watching TV or Playing Video Games by Kids – Pros and Cons

What is better? Watching TV or playing video games? What is common between them is that one is watched on a screen and another played on a screen. In a sense, both are equally harmful because kids have to strain their eyes which in turn will affect their vision.

There may not be many similarities between a kid who watches programs on TV and one who plays video games. If a kid watches TV programs, he or she usually watches cartoons. Cartoons do not give you any fruitful information on anything at all. A kid sits idly in front of a TV set for hours together with its eyes glued to it. But, with regard to video games, the kid actually gets to do something – moves its hand or does something to the character. So it is more immersive but more immersive does not mean better all the way. Though sometimes video games are violent, equally violent are some scenes in which superheroes are fighting against people. There are some benefits to playing video games. There are ratings in video games which restrict children from buying violent video games. But with respect to TV programs, they are not rated and are fully open to children. Now, companies such as LG and Samsung are coming up with new and innovative ways to block a channel or block the children from watching a channel that is harmful to them or bad for them. Ratings exist because the more mature you become, the less adaptive you are and you do not want to experiment with that in real life. That is why ratings work. With regard to TV programs, there are no ratings. It means any child can watch any kind of violent scenes.

In both cases—video games and TV programs—children have to sit upright. Instead, if they are sitting in a wrong posture, it may be bad for their spine.

Both of them have their pluses and minuses. Both violent video games and TV programs are bad in their own right. But if children abide by the PEGI ratings, which are restrictions that require them to be of a certain age to purchase a video game, it will be fine. If we consider normal video games and TV programs that are appropriate for a child's age, then it is good.

41
The Huawei Ban Explained

Huawei, a Chinese electronics manufacturer, has been banned from transacting business with any American companies. This has some huge implications as Huawei depends on many American companies to make hardware and software for it.

The two main reasons for the ban are strained trade relations with China and security concerns. The USA had a suspicion that Huawei was collecting personal data from its users and sharing it with the Chinese government. It is a security concern. Huawei also makes networking equipment that is used by many companies based in the USA. When asked in an interview about the situation, the US President Donald Trump said, "Huawei is something that's very dangerous. You look at what they've done from a security standpoint, from a military standpoint, it's very dangerous. So it's possible that Huawei even would be included in some kind of a trade deal. If we made a deal, I could imagine Huawei being possibly included in some form, some part of a trade deal." The US President's assertion has caused many problems for Huawei. It seems that Huawei has been preparing for this for months now.

Google is a US company and Huawei makes Android phones. Android is a Google company. So, now Huawei will probably not get new updates. Huawei has fired back with making its own operating system called "HongMeng OS." The Play Store is also owned by Google. Huawei probably cannot use it. It will be hard to sell its phones without the Android OS and Play Store.

The Samsung Galaxy Fold failed and the next promising folding phone was the Huawei Mate X. It will be very hard to convince people to pay $1000 for a folding phone without Android and Play Store. The US companies make a lot of hardware that Huawei uses. Huawei has to make those parts in-house or have a stockpile of those parts for the future. This will be very expensive. ZTE also faced a similar situation. As a result, it was crippled and went out of business. But Huawei is a huge company and so they may recover but only time will tell.

(Source: 41.1)

MATHEMATICS

42
Pi (π – The Magical Number)

Pi has been a very important number for so many centuries. Let us see its history. The ancient Babylonian civilization has contributed to the history of pi. The Babylonians first estimated 3 to be the value of pi. One Babylonian tablet dating back to the period between 1900 and 1680 B.C. has recorded a value of 3.125 for pi which is closer to the accurate estimate.

The ancient Egyptians calculated the area of a circle by a formula that gave pi the approximate value of 3.1605 in 1650 B.C. A very renowned Greek mathematician, Archimedes, did one of the earliest calculations of pi. After the Babylonians, Archimedes worked on pi from very early days. He approximated and calculated the area of a circle using the Pythagorean Theorem. He used a polygon with thousands of sides for calculating the value of pi.

Zu Chongzhi, a brilliant Chinese mathematician and astronomer, used a similar method as Archimedes. He would not have been familiar with Archimedes' method as they had lived during different periods in different places. Zu Chongzhi's book has been lost and so little is known about his work.

Many great mathematicians throughout history have contributed to finding the value of pi. Nowadays, pi is calculated using computers. Currently, we have calculated 2,700,000,000,000 (2.7 trillion) decimal places of pi which is nearest to the accurate. Pi has an infinite number of decimals as it is an irrational.

There have been many iconic numbers that are very important in mathematics like "e" known as Euler's number and "I" imaginary number. Most of the iconic numbers are from one-time mathematical superpowers of the world – Persia, Greece, India and China. It is undoubtful that pi is the most iconic number in all of mathematics.

Akira Haraguchi, a Japanese, memorized 1,00,000 digits of pi. It is amazing and it demonstrates his astounding memory.

The value of pi keeps on extending; it gets closer and closer to the final answer but it never gets the exact value. We are always close but far from the exact value. It cannot be expressed as a fraction; its decimal representation never ends. The fraction 22/7 is commonly used to approximate π, but actual π has an infinite number of digits. But in school, you may have learned the value of pi as 22/7 which is only an approximation.

Pi is commonly defined as the ratio of a circle's circumference to its diameter and so it is equal to circumference by the diameter.

(Source: 42.1)

43

Why Can't We Divide Any Number by Zero?

N/zero =?

In school, we are taught that we should not and cannot divide any number by zero. Why is it so? Before going into the details, let us analyze the mathematical background. Let us get into very simple mathematics. Division is repeated subtraction. Also, it is equal distribution.

Let us take number 1, for example.

   1-0 = 1

You can do it almost forever. It is shown below:

   1-0-0-0-0-0-0-0-0-0-0-0-0......................∞ (infinity)

You can do that forever and ever, that is, infinity (∞). Basically, the answer is supposed to be infinity.

Let us approach the same problem from a different angle.

So we are going to see what 1 over 1 is.

   1 over 1=1

   1 over 0.1=10

   1 over 0.01=100

   1 over 0.001=1000 and so on.

So after infinite series, it just decreases to 1 over 0.

The answer to 1 over 0 is infinity.

So far, the answer has been infinity, but we have been taught that we should not divide any number by 0.

We did it with a positive number.

Now let us see it on the negative side.

So let us do it with negative numbers.

>1 over -1 = -1

>1 over -0.1 = -10

>1 over -0.01 = -100 and so on.

If we keep doing, it will approach minus infinity (-∞). Let me prove that it is not.

Now, look at this.

>1-0-0-0-0-0-0...∞

>1 over 1 = 1

>1 over 0.1 = 10

>1 over 0 = ∞

>1 over -1 = -1

>1 over -0.1= -10

Let us now do it with 2.

>2 over 1= 2

>2 over 0.1=20

>2 over 0.01=200 and so on.

Finally,

2 over 0 = ∞

That means 1 = 2.

But common sense tells us that 1 is not equal to 2. We do not need to know big calculations to know that 1 is not equal to 2.

By dividing any number by 0 and finding that the answer is infinity (∞), it proposes that 1 is equal to 2 which is not true. On the positive side, we get infinity (∞); on the negative side, we get minus infinity (–∞). Even with 2, 3, 4 which are different values, it still adds up to infinity which would make this problem wrong. So, any number divided by 0 is called an undefined function. We call it undefined not because mathematicians are lazy, but because this is an unpredictable function that suggests 1 cannot be divided by 0. It gives us same answers in different circumstances which are not valid. This function is unpredictable and does not abide by the laws of mathematics. That is why we cannot divide any number by 0.

(Source: 43.1)

ENGLISH FOR SCIENCE AND TECHNOLOGY

44

The Impact of Science and Technology YouTube Videos and English TV Programs on My Speaking Skills – Part I

(My Journey Through the World of English)

I will be explaining how I acquired English-speaking skills by recounting some milestones I have crossed during my journey through the world of English. I have acquired proficiency in English, and specifically technical English due to two factors: watching science and technology YouTube videos and English channels that focus on science and technology.

First, let me discuss in detail how my exposure to YouTube videos influenced my English. Except for my school hours, the rest of the time I watch the videos of native speakers of English. While watching the videos, I observe how they speak and then I try to speak like them. When you just get immersed in it, it will be much easier for you to learn English. That is how it has become natural and normal for me to use the language with quite ease and comfort.

Many people speak English on YouTube, but I urge you to listen to the English of native speakers. Only then can you learn English—pronunciation, stress, intonation—properly.

There is a YouTube channel called "Good Mythical Morning" owned by two persons named Rhett and Link. I started watching their channel three years ago and since then I have been watching it. Their YouTube videos inspired me to speak English well, and consequently, I created my own YouTube channel named "Frentran." I started speaking a lot of English. So far, I have uploaded 53 videos on different subjects, mainly on science and technology, on my YouTube channel Frentran. I still keep on learning the nuances of the English language and science and technical terms while watching the science and technology YouTube videos of Linus Tech Tips, TechLinked, Austin Evans, CrashCourse, SmarterEveryDay, CHM Tech, TED-ED, Unbox Therapy, Numberphile, The Infographics Show, Techquickie, Veritasium, Marques Brownlee, This is, Snazzy Labs and TechAltar.

As to my English, even the school did not influence me as YouTube did. YouTube influenced me in a good, powerful and strong way. I can now speak English as good and as fluent as I speak my mother tongue Tamil. I want to stress the fact that it is how much attention you pay really matters. In other words, the more attention you pay, the more you can grasp from the videos on science and technology.

Tom Scott is a British YouTuber who makes videos on a wide range of topics from facts to fiction. Apart from learning good English, you can also learn some scientific facts and terms from his videos.

Captain Disillusion is a YouTuber born in Latvia and raised in America. He makes videos on digital effects. His videos are very well made, and he has flawless intonation. Apart from that, you can learn many words pertaining to digital effects.

Drew Binsky is an American YouTuber who makes videos on traveling. He speaks American English amazingly. Besides learning English, you can also learn about the cultures of different countries.

Matt D'Avella is also an American YouTuber who makes videos on minimalism and how to be successful in life. You can learn a wide range of words from him. His frequent use of difficult words helps you learn their pronunciation easier and quicker.

There are many more YouTubers who influenced me to learn a lot of scientific terms. As a result, I am able to speak and write better on scientific topics. But for want of space, I have listed the most influential ones.

Second, let me explain how watching English TV channels helped me improve my English-speaking skills enormously. Immediately after returning from school, I would sit down and watch only TV programs in English. I watch FYI TV 18 and it telecasts an American TV show called "Tiny House World" where they show how to build tiny houses for a small budget. The sellers and prospective buyers converse in English naturally. So, we are able to hear their casual way of speaking American English.

There is another show on this TV called "Man vs Child–Chef Showdown." And in this show adults and kids participate in cooking contests. By watching this program, we can get a pretty good knowledge of cooking different dishes besides learning English.

There is also another TV channel called "History TV 18" and it telecasts an amazing history show called "Top 10" in which they exhibit ten ancient things of historical significance, such as ancient devices, weapons and pottery. We can learn English in an interesting, amazing and a good fun way. There is another program called "Pawn Stars" in which people bring artifacts to a pawnshop and the shop owners appraise their value and buy them. We can learn a lot of English through this show.

There is another TV channel called "Colours Infinity." It telecasts a show called "America's Got Talent." It is an American TV show where you can watch people from all over the world showcasing their talents and the judges Simon Cowell, Mel B, Heidi Klum and Howie Mandel express their viewpoints and feelings. We can also learn how to comment on a performance. It is really a useful channel.

"America Ninja Warrior" is a program on AXN for the sports lovers. It is a competition where the contestants try to get through an obstacle course. You can also learn very good English from it.

Keep watching a lot of YouTube videos of native speakers of English and a lot of English channels. By doing so, you can learn to speak English fluently and faultlessly.

45

The Impact of Science and Technology YouTube Videos and English TV Programs on My Speaking Skills – Part II

Speaking in English fluently and faultlessly is an art. Non-native English speakers can learn this art by learning language skills. You need to improve your listening, speaking, reading and writing skills, and specifically listening and speaking skills to speak fluently and flawlessly.

Here are some important tips for you.

Tip 1: Believe in Yourself

First, you should have confidence in yourself. You should have the conviction that you can speak in English fluently and correctly. There are two important aspects of any language: fluency and accuracy. First, you should strive for acquiring fluency. Second, you can concentrate on the accuracy of the language. Mastering a language is easy, provided you have the motivation, determination, conviction and intensive practice.

When you start speaking in English, people may tell you that your English is unintelligible. But, you should not lose hope because you are in the process of learning. And making mistakes is part and parcel of any learning experience. You should take the criticism positively and move forward. It should not hurt you because it is said so to help you put yourself on the right track. The next important step is practice. Practice, practice, practice. Practice plays a pivotal role in learning proper English. The more you practice speaking in English, the better speaker you are. The English language is funny, complicated and even sometimes bizarre.

Tip 2: Do It Your Way

The English language is spoken with different accents. But, the most important ones are British accent and American accent. There are variations of the British accent, such as Welsh, Scottish and Northern Irish. However, these accents are widely different from the main British accent.

American accent also has variations. The various accents are due to different regions and states. But, they are not very different from the common American accent.

You can speak in whatever accent you feel most comfortable with and use any variation of a word that you want to use, as far as you are correct. If you want to learn English well, you have to learn to speak it often. I have learned most of my English from watching a lot of science and technology YouTube videos of native speakers of English and English TV programs for the past two years. Everybody has their own way of doing things. You should have your own way of speaking English as far as your grammar and pronunciation are correct. If you intend to go to the USA for higher studies or a job, you need to learn to speak with an American accent.

Tip 3: Expand Your Vocabulary

It is very important that you should learn a lot of scientific terms. Why should you keep learning new words? Even an experienced native English speaker does not know all the words in the English language. We cannot know all the words in English. Moreover, every year hundreds of scientific words are added to the dictionary. Yet if you learn more words by watching science and technology YouTube videos and English TV programs, you can express your thoughts and ideas, opinions and views very effectively. If you come across unknown and unfamiliar words, you can learn the meanings of the words in context or look them up in a dictionary to find the meanings and usage of them.

Tip 4: Do Not Assume Pronunciation of Words

You should never assume a word's pronunciation by looking at its spelling because there is no correlation between the spelling and pronunciation of most of the words in English. For example, take the word "chateau" which means "castle." You may assume that it is pronounced "chatoo," but the correct pronunciation is "shatoe." That is why you should not assume the pronunciation of words by considering only the spellings. You may get the pronunciation wrong.

Tip 5: Listen to Native English Speakers

Listen to native English speakers because they have been speaking English since their birth as English is their mother tongue. If you keep listening to native speakers, the right way to speak English will become engrained in your mind. That is how I got my English.

I wish you the best on your journey to learn English the right way.

References

1.1 The History of Paper. (2019). Retrieved from **https://www.papertrading.com/prod01.htm**

1.2 The History of Paper. (2019). Retrieved from **https://www.papertrading.com/prod01.htm**

1.3 Tyre. (2019). Retrieved from **https://en.wikipedia.org/wiki/Tyre**

1.4 Smartphone. (2019). Retrieved from **https://en.wikipedia.org/wiki/Smartphone**

2.1 Pogue, D. (2019). Breaking the Myth of Megapixels. Retrieved from **https://www.nytimes.com/2007/02/08/technology/08pogue.html?pagewanted=all&_r=2&**

2.2 (2019). Retrieved from **https://www.fcc.gov/consumers/guides/wireless-devices-gas-stations**

2.3 O'Connor, A. (2019). Microwave Ovens - Safety - Radiation. Retrieved from **https://www.nytimes.com/2007/07/10/health/10real.html**

2.4 11 Common Myths About the Technology You Use Every Day. (2019). Retrieved from **https://www.huffingtonpost.in/2014/09/15/tech-myths_n_5791350.html**

2.5 Do television and electronic games predict children's psychosocial adjustment? Longitudinal research using the UK Millennium Cohort Study. (2019). Retrieved from **https://adc.bmj.com/content/98/5/341**

3.1 Microplastics in Seafood and the Implications for Human Health. Retrieved from **https://www.ncbi.nlm.nih.gov/pmc/articles/PMC6132564**

3.2 A Spoon You Can Eat Is a Tasty Alternative to Plastic Waste. (2016). [Video]. Retrieved from **https://www.youtube.com/watch?v=r4Cc5zmy0eY**

3.3 This biodegradable water bottle could replace plastic. (2017). [Video]. Retrieved from **https://www.youtube.com/watch?v=rVLx12Ebryw**

3.4 AFP News Agency. (2017). [Video]. Retrieved from **https://www.youtube.com/watch?v=dXklBP53VT4**

3.5 Motherboard. (2015). Fungus: The Plastic of the Future [Video]. Retrieved from **https://www.youtube.com/watch?v=jnMXH5TqqG8**

3.6 CNBC. (2017). The Ooho Is an Edible, Biodegradable Water Bottle with a Jelly-Like Skin [Video]. Retrieved from **https://www.youtube.com/watch?v=Pj6Q-YCcA3s**

4.1 50 Awesome Facts (About Everything). (2019). Retrieved from **http://mentalfloss.com/article/58321/50-awesome-facts-about-everything**

5.1 Bramble Cay melomys. (2019). Retrieved from **https://en.wikipedia.org/wiki/Bramble_Cay_melomys**

5.2 Bulldog Rat. Retrieved from **https://en.wikipedia.org/wiki/Bulldog_rat**

5.3 De-extinction. (2019). Retrieved from **https://en.wikipedia.org/wiki/De-extinction**

6.1 Martin, S. (2016). The 10 weirdest planets to have been discovered so far. Retrieved from **https://www.express.co.uk/news/science/643662/The-10-weirdest-planets-to-have-been-discovered-so-far**

7.1 Michael Stevens, Vsauce. (2012). How Hot Can It Get? [Video]. Retrieved from **https://www.youtube.com/watch?v=4fuHzC9aTik**

8.1 Wright Flyer. (2019). Retrieved from **https://en.wikipedia.org/wiki/Wright_Flyer**

8.2 Learn Engineering. (2018). How do Airplanes fly? [Video]. Retrieved from **https://www.youtube.com/watch?v=F077WDnB8P8**

8.3 Howcast. (2011). What's in the Cockpit? | Flying Lessons [Video]. Retrieved from **https://www.youtube.com/watch?v=4r2LqrlpkJc**

8.4 Mentour Pilot. (2018). Why do aircraft store fuel in the wings? [Video]. Retrieved from **https://www.youtube.com/watch?v=VpzUbYex4dg**

9.1 History of Helicopters. (2019). Retrieved from **https://en.wikipedia.org/wiki/Helicopter#History**

9.2 Garden, H. (2019). How Helicopters Work. Retrieved from **https://science.howstuffworks.com/transport/flight/modern/helicopter.htm**

10.1 Veritasium. (2017). World's First Car! [Video]. Retrieved from **https://www.youtube.com/watch?v=DL_mJeb6O04**

11.1 List of Lamborghini automobiles. Retrieved from **https://en.wikipedia.org/wiki/List_of_Lamborghini_automobiles**

11.2 Lamborghini. Retrieved from **https://en.wikipedia.org/wiki/Lamborghini**

12.1 Radioactivity : Uranium 238 and 235. Retrieved from **http://www.radioactivity.eu.com/site/pages/Uranium_238_235.htm**

12.2 Pros and Cons of Nuclear Energy. (2019). Retrieved from **https://www.conserve-energy-future.com/pros-and-cons-of-nuclear-energy.php**

13.1 Explorer 1. (2019). Retrieved from **https://en.wikipedia.org/wiki/Explorer_1**

13.2 M. Harland, D. Chandra X-ray Observatory | United States satellite. Retrieved from **https://www.britannica.com/topic/Chandra-X-Ray-Observatory**

13.3 About the Hubble Space Telescope. (2017). Retrieved from **https://www.nasa.gov/content/goddard/nasa-refresher-on-hubble-facts-for-the-25th-anniversary**

13.4 Apollo 13. (2019). Retrieved from **https://en.wikipedia.org/wiki/Apollo_13**

13.5 Mercury-Redstone 3. (2019). Retrieved from **https://en.wikipedia.org/wiki/Mercury-Redstone_3**

14.1 Artemis Moon Program Advances – The Story So Far. Retrieved from **https://www.nasa.gov/artemis-moon-program-advances**

14.2 NASA. (2019). Our Next Lunar Landings [Video]. Retrieved from **https://www.youtube.com/watch?v=qODDdqK9rL4**

14.3 NASA. (2019). We are Going [Video]. Retrieved from **https://www.youtube.com/watch?v=vl6jn-DdafM**

15.1 SciShow. (2012). Absolute Zero: Absolute Awesome [Video]. Retrieved from **https://www.youtube.com/watch?v=TNUDBdv3jWI**

16.1 Matthew Patrick (MatPat). (2016). Neurons and Memory formation [Video]. Retrieved from **https://www.youtube.com/watch?v=1fV3jafglNo**

17.1 A. Abbott, E. (1884). Flatland: A Romance of Many Dimensions (p. 7). London: Seeley & Co.

17.2 Kurzgesagt-In a Nutshell. (2018). Wormholes Explained – Breaking Spacetime[Video]. Retrieved from **https://www.youtube.com/watch?v=9P6rdqiybaw**

18.1 Seeker. (2017). This is the only place in the universe where Antimatter can survive[Video]. Retrieved from **https://www.youtube.com/watch?v=XD8Q3Mb1Q4I**

19.1 Albert Einstein. (2019). Retrieved from **https://en.wikipedia.org/wiki/Albert_Einstein**

19.2 TED. (2015). Einstein's miracle year - Larry Lagerstrom [Video]. Retrieved from **https://www.youtube.com/watch?v=91XI7M9l3no**

21.1 Retrieved from **https://www.who.int/**

23.1 KFC "Grab a Bucket of Chicken". [Video]. Retrieved from **https://www.youtube.com/watch?v=lrcuIGypxDg**

26.1 [Ebook]. Retrieved from **https://www.dartmouth.edu/~dcomin/index_files/wsj.pdf**

26.2 Tim Berners-Lee. Retrieved from **https://en.wikipedia.org/wiki/Tim_Berners-Lee**

28.1 TED. (2018). How AI can save our humanity | Kai-Fu Lee [Video]. Retrieved from **https://www.youtube.com/watch?v=ajGgd9Ld-Wc**

28.2 Boyinaband. (2017). Why AI will probably kill us all. [Video]. Retrieved from **https://www.youtube.com/watch?v=SPAmbUZ9UKk**

31.1 Computerphile. (2013). How NOT to Store Passwords! - Computerphile [Video]. Retrieved from **https://www.youtube.com/watch?v=8ZtInClXe1Q**

32.1 Menegus, B. Retrieved from **https://gizmodo.com/its-illegal-to-possess-or-distribute-this-huge-number-1774473790**

33.1 Bill Gates. (2019). Retrieved from **https://en.wikipedia.org/wiki/Bill_Gates**

33.2 Bill & Melinda Gates Foundation. Retrieved from **https://www.gatesfoundation.org/**

33.3 Steve Jobs. Retrieved from **https://en.wikipedia.org/wiki/Steve_Jobs**

33.4 Jeff Bezos. Retrieved from **https://en.wikipedia.org/wiki/Jeff_Bezos**

33.5 Mark Zuckerberg. Retrieved from **https://en.wikipedia.org/wiki/Mark_Zuckerberg**

33.6 Singel, R., Baker-Whitcomb, A., Harrison, S., Finley, K., Barber, G., Tiku, N., & Barber, G. Mark Zuckerberg: I Donated to Open Source, Facebook Competitor. Retrieved from **https://www.wired.com/2010/05/zuckerberg-interview/**

33.7 Elon Musk. Retrieved from **https://en.wikipedia.org/wiki/Elon_Musk**

34.1 Microsoft. Retrieved from **https://en.wikipedia.org/wiki/Microsoft**

35.1 Adobe Photoshop. Retrieved from **https://en.wikipedia.org/wiki/Adobe_Photoshop**

38.1 Video games - Wikipedia. Retrieved from **https://en.wikipedia.org/wiki/Video_game**

41.1 Brownlee, M. (2019). The Huawei Ban: Explained! [Video]. Retrieved from **https://www.youtube.com/watch?v=qZGpmWrVSaU**

42.1 A Brief History of Pi (π) | Exploratorium. Retrieved from **https://www.exploratorium.edu/pi/history-of-pi**

43.1 Woo, E. (2014). Dividing by zero? [Video]. Retrieved from **https://www.youtube.com/watch?v=J2z5uzqxJNU**

Links for Pictures

3. **Five Alternative Materials to Plastic**
   Mycelium Chair:
   https://commons.wikimedia.org/wiki/File:Mycelium_Chair.jpg
4. **Amazing Facts That Will Teach You Something New**
   Gecko Feet:
   https://en.wikipedia.org/wiki/File:Gecko_foot_on_glass.JPG
   Nile Crocodile:
   https://en.wikipedia.org/wiki/Nile_crocodile#/media/File:NileCrocodile.jpg
   Armadillo:
   https://en.wikipedia.org/wiki/Armadillo#/media/File:Nine-banded_Armadillo.jpg
5. **How Can We Resurrect Extinct Animals?**
   Rat:
   https://www.businessinsider.in/Meet-the-first-species-to-go-extinct-because-of-climate-change-it-was-tiny-cute-and-fluffy/articleshow/68104580.cms
   (taken from labeled for reuse)

Bulldog Rat:

https://en.wikipedia.org/wiki/Bulldog_rat#/media/File:Rattus_nativitatis.jpg

Dolly the Sheep:

https://en.wikipedia.org/wiki/Dolly_(sheep)#/media/File:Dolly_face_closeup.jpg

7. **What is the Hottest Object in the Universe?**

   Eta Carinae:

   https://upload.wikimedia.org/wikipedia/commons/thumb/f/fc/Eta_Carinae.jpg/280px-Eta_Carinae.jpg

8. **How Do Airplanes Fly?**

   Wright Flyer:

   https://commons.wikimedia.org/wiki/File:Kitty_hawk_gross.jpg

   Airfoil:

   https://commons.wikimedia.org/wiki/File:Wing_profile_nomenclature.svg

   Airplane Cockpit:

   https://ja.wikipedia.org/wiki/%E3%82%A8%E3%82%A2%E3%83%90%E3%82%B9A300#/media/%E3%83%95%E3%82%A1%E3%82%A4%E3%83%AB:Airbus_A300B2-103,_Novespace_(CNES)_AN1993670.jpg (taken from labeled for reuse with modification)

9. **How Do Helicopters Fly?**

   Igor Sikorsky:

   https://upload.wikimedia.org/wikipedia/commons/thumb/4/4d/Sikorsky%2C_Igor.jpg/220px-Sikorsky%2C_Igor.jpg

Parts of a helicopter:

https://commons.wikimedia.org/wiki/File:
Helicopter_expanded_view.svg

Helicopter:

https://commons.wikimedia.org/wiki/Helicopter#/
media/File:Bell407_GNU-FDL.jpg

Tail Rotor:

https://commons.wikimedia.org/w/index.
php?search=tail+rotor&title=Special%3ASearch &go=G
o&ns0=1&ns6=1&ns12=1&ns14=1&ns100=1&ns106=1#/
media/File:Tail_rotor_mechanism_CH-53G.jpg

Cockpit:

https://en.wikipedia.org/wiki/File:Helicopter_
controls_layout.svg

10. Mankind's Quest for Speed

Benz Patent Motor Car:

https://en.wikipedia.org/wiki/File:1885Benz.jpg

Blue Flame:

https://commons.wikimedia.org/wiki/File:
Goodwood2007-121_The_Blue_Flame.jpg

Crash Test Dummies:

https://commons.wikimedia.org/wiki/
File:CEP1710-58.jpg

11. Automobile Industry

Ferrari F40:

https://commons.wikimedia.org/wiki/File:F40_
Ferrari_20090509.jpg

Lamborghini Aventador:

https://commons.wikimedia.org/wiki/File:Lamborghini_Aventador_(5488245669).jpg

Bugatti Veyron:

https://commons.wikimedia.org/wiki/File:Bugatti_veyron_in_Tokyo.jpg

Dodge Tomahawk:

https://commons.wikimedia.org/wiki/File:Dodge_Tomahawk_Concept.jpg

Tesla Roadster 2020:

https://commons.wikimedia.org/wiki/File:Tesla_Roadster_2.0_(47619421652).jpg

Pagani Zonda:

https://commons.wikimedia.org/wiki/File:Red_Pagani_Zonda_Roadster_in_Monaco_2012.jpg

12. Nuclear Energy

Nuclear Fission:

https: //commons.wikimedia.org/ w/index.php?sort=relevance&search=nuclear+fission&title=Special:Search&profile=advanced&fulltext=1&advancedSearch-current=%7B%7D&ns0=1&ns6=1&ns12=1&ns14=-1&ns100=1&ns106=1#/media/File:Nuclear_fission.svg

Nuclear Reactor:

https://libreshot.com/nuclear-power-plant/(taken from labeled for reuse)

Thorium:

https://sr.m.wikipedia.org/wiki/ %D0%94%D0%B0%D1%82%D0%BE%D1%82%D0%B5%D0%BA%D0%B0: Thorium-1.jpg

Uranium:

https://commons.wikimedia.org/wiki/ File:Uranium_ore_square.jpg

13. The Achievements of NASA

Explorer 1:

https://www.flickr.com/photos/11304375@N07/ 6794065325 (taken from labeled for reuse)

Chandra X-ray observatory:

https://commons.wikimedia.org/wiki/ File:Chandra_X-ray_space_observatory_-_ LightpathQ202.jpg

Hubble Space Telescope:

https://en.wikipedia.org/wiki/Hubble_Space_ Telescope#/media/File:HST-SM4.jpeg

Pioneer 10:

https://upload.wikimedia.org/wikipedia/ commons/f/f0/An_artist%27s_impression_of_a_ Pioneer_spacecraft_on_its_way_to_interstellar_ space.jpg

Apollo 13 Service Module:

https://upload.wikimedia.org/wikipedia/ commons/4/43/Apollo_13_Service_Module_-_AS13- 59-8500_%28cropped%29.jpg

Freedom 7:

https://en.wikipedia.org/wiki/Mercury-Redstone_3#/media/File:Mercury-Redstone_3_Launch_MSFC-6100884.jpg

International Space Station:

https://upload.wikimedia.org/wikipedia/commons/3/32/ISS_after_STS-119_in_March_2009_1.jpg

Apollo 11:

https://en.wikipedia.org/wiki/Apollo_11#/media/File:Apollo_11_Launch_-_GPN-2000-000630.jpg

Armstrong:

https://en.wikipedia.org/wiki/Apollo_11#/media/File:Armstrong_on_Moon_(As11-40-5886)_(cropped).jpg

15. Absolute Zero

Boomerang Nebula:

https://upload.wikimedia.org/wikipedia/commons/thumb/b/b2/Boomerang_nebula.jpg/220px-Boomerang_nebula.jpg

16. How Does Our Brain Store Memories?

Neuron:

https://www.maxpixel.net/Neurons-Brain-Structure-Network-Brain-Cells-Brain-440660 (taken from labeled for reuse)

Hippocampus:

https://upload.wikimedia.org/wikipedia/commons/2/2e/Gray739-emphasizing-hippocampus.png

Pre-frontal Cortex:

https://commons.wikimedia.org/wiki/File:
Prefrontal_cortex_of_the_brain.png

17. Higher Dimensions

4D Cube:

https://upload.wikimedia.org/wikipedia/commons
/5/55/8-cell-simple.gif

Wormhole:

https://commons.wikimedia.org/wiki/File:
Wormhole.png

Negative Mass:

https://upload.wikimedia.org/wikipedia/commons/
thumb/c/ca/Approximation_newtonienne_masse.
svg/200px-Approximation_newtonienne_masse.svg.png

19. The Story of a Genius: Albert Einstein

Albert Einstein:

https://upload.wikimedia.org/wikipedia/commons/
thumb/3/3e/Einstein_1921_by_F_Schmutzer_-_
restoration.jpg/220px-Einstein_1921_by_F_
Schmutzer_-_restoration.jpg

Hiroshima & Nagasaki Bomb:

https://en.wikipedia.org/wiki/Fat_Man#/media/
File:Nagasakibomb.jpg

20. Global Warming

Tuvalu:

https://commons.wikimedia.org/wiki/File: Tuvalu_
Inaba-21.jpg

22. Fast food or Healthy Food – What Should We Eat?

Food Pyramid:

https://commons.wikimedia.org/wiki/File:Harvard_food_pyramid.png

23. Does Advertising Influence the Food Choices of Children?

Burger:

https://pixabay.com/photos/burger-hamburger-bbq-cheeseburger-3962996/ (taken from labeled for reuse)

24. Parts of a Computer and Their Functions

Computer Case:

https://en.wikipedia.org/wiki/File:AVADirect-Custom-X99-Intel-Core-i7-gaming-cpu.png

Motherboard:

https://www.flickr.com/photos/126089327@N04/14898499521 (taken from labeled for reuse)

Processor:

https://commons.wikimedia.org/wiki/File:Intel_CPU_Core_i7_6700K_Skylake_perspective.jpg

Hard Drive:

https://upload.wikimedia.org/wikipedia/commons/5/5f/Airy_by_CnMemory%2C_external_hard_disk-3187.jpg

RAM:

https://commons.wikimedia.org/wiki/File:2014_Corsair_Dominator_Platinum_2x4GB,_1866MHz.JPG

26. Interesting Facts About Computers

UNIVAC I:

https://commons.wikimedia.org/wiki/File:Museum_of_Science,_Boston,_MA_-_IMG_3163.JPG

First Mouse:

https://www.flickr.com/photos/41176169@N00/2642494671 (taken from labeled for reuse)

Tim Berners Lee:

https://en.wikipedia.org/wiki/Tim_Berners-Lee#/media/File:Sir_Tim_Berners-Lee_(cropped).jpg

27. Computers and Technology – Questions and Answers

Xbox One:

https://commons.wikimedia.org/wiki/File:Xbox_One_S_All-Digital_Edition.png

Pop Filter:

https://commons.wikimedia.org/wiki/File:GaugeInc_PopFilter.png

28. Artificial Intelligence (AI)

Artificial Intelligence:

https://www.flickr.com/photos/mikemacmarketing/30212411048 (taken from labeled for reuse)

Self-Driving Car:

https://commons.wikimedia.org/wiki/File:Inside_the_Google_RoboCar_today_with_PlanetLabs.jpghttps://commons.wikimedia.org/wiki/File:Inside_the_Google_RoboCar_today_with_PlanetLabs.jpg

Robot:

https://upload.wikimedia.org/wikipedia/commons/thumb/6/6c/Atlas_from_boston_dynamics.jpg/220px-Atlas_from_boston_dynamics.jpg

31. How Do Websites Store Your Password?

Encryption Diagram:

https://www.flickr.com/photos/166102838@N03/31024015727 (taken from labeled for reuse)

Hashing Diagram:

https://commons.wikimedia.org/wiki/File:CPT-Hashing-Password-Bad.svg

33. The Most Influential Innovators and Their History

Bill Gates:

https://en.wikipedia.org/wiki/Bill_Gates#/media/File:Bill_Gates_July_2014.jpg

Steve Jobs:

https://en.wikipedia.org/wiki/Steve_Jobs#/media/File:Steve_Jobs_Headshot_2010-CROP_(cropped_2).jpg

Jeff Bezos:

https://en.wikipedia.org/wiki/Jeff_Bezos#/media/File:Jeff_Bezos_at_Amazon_Spheres_Grand_Opening_in_Seattle_-_2018_(39074799225)_(cropped).jpg

Mark Zuckerberg:

https://en.wikipedia.org/wiki/Mark_Zuckerberg#/media/File:Mark_Zuckerberg_F8_2018_Keynote_(cropped_2).jpg

Elon Musk:

https://en.wikipedia.org/wiki/Elon_Musk#/media/File:Elon_Musk_Royal_Society.jpg

34. The History of Microsoft

Altair 8800:

https://en.wikipedia.org/wiki/Altair_8800#/media/File:Altair_8800_Computer.jpg

MS-DOS:

https://en.wikipedia.org/wiki/MS-DOS#/media/File:Msdos-icon.png

Windows Phone:

https://www.flickr.com/photos/125207874@N04/14536343286 (taken from labeled for reuse)

Steve Ballmer:

https://en.wikipedia.org/wiki/Steve_Ballmer#/media/File:Steve_Ballmer_2014.jpg

Satya Nadella:

https://commons.wikimedia.org/wiki/File:Satya_smiling-print.jpg

HoloLens 2:

https://commons.wikimedia.org/wiki/File:HoloLens_2.jpeg

35. The Most Innovative Software and Websites

Facebook logo:

https://upload.wikimedia.org/wikipedia/commons/thumb/c/cd/Facebook_logo_%28square%29.png/600px-Facebook_logo_%28square%29.png

Adobe Photoshop logo:

https://commons.wikimedia.org/wiki/Category:Adobe_Photoshop#/media/File:Adobe_Photoshop_CC_icon.svg

Google logo:

https://commons.wikimedia.org/wiki/Google#/media/File:GoogleLogoSept12015.png

YouTube logo:

https://commons.wikimedia.org/wiki/File:YouTube_Logo_2017.svg

WhatsApp logo:

https://en.m.wikipedia.org/wiki/File:WhatsApp.svg

36. Cloud Computing

Cloud Computing Diagram:

https://upload.wikimedia.org/wikipedia/commons/2/27/Cloud_applications.jpg

Amazon Web Services:

https://upload.wikimedia.org/wikipedia/commons/9/93/Amazon_Web_Services_Logo.svg

GeForce Now:

https://www.flickr.com/photos/gbpublic/21846230641/ (taken from labeled for reuse)

www.ingramcontent.com/pod-product-compliance
Lightning Source LLC
Chambersburg PA
CBHW030930180526
45163CB00002B/516